Eurotherm 105: Computational Thermal Radiation in Participating Media V

Journal of Physics: Conference Series Volume 676

Albi, France
1 – 3 April 2015

Editors:

Mouna EL Hafi
Richard Fournier
Denis Lemonnier
Paul Lybaert
Nevin Selcuk

ISBN: 978-1-5108-1966-5
ISSN: 1742-6588

Printed from e-media with permission by:

Curran Associates, Inc.
57 Morehouse Lane
Red Hook, NY 12571

CURRAN ASSOCIATES INC. proceedings.com

Some format issues inherent in the e-media version may also appear in this print version.

Printed by Curran Associates, Inc. (2016)

For permission requests, please contact the Institute of Physics
at the address below.

Institute of Physics
Dirac House, Temple Back
Bristol BS1 6BE UK

Phone: 44 1 17 929 7481
Fax: 44 1 17 920 0979

techtracking@iop.org

Additional copies of this publication are available from:

Curran Associates, Inc.
57 Morehouse Lane
Red Hook, NY 12571 USA
Phone: 845-758-0400
Fax: 845-758-2634
Email: curran@proceedings.com
Web: www.proceedings.com

Table of contents

Volume 676

Eurotherm Conference 105: Computational Thermal Radiation in Participating Media V
1–3 April 2015, Albi, France

Accepted papers received: 2 December 2015
Published online: 26 January 2016

Preface

Papers

012024

Meshless method for solving coupled radiative and conductive heat transfer in refractive index medium Cheng-An Wang, Hamou Sadat and Jian-Yu Tan pg. 161

Journal of Physics: Conference Series **676** (2016) 011001 doi:10.1088/1742-6596/676/1/011001

Eurotherm Conference No. 105: Computational Thermal Radiation in Participating Media V

Mouna EL Hafi, Richard Fournier, Denis Lemonnier, Paul Lybaert and Nevin Selçuk

This volume of *Journal of Physics: Conference Series* is based on papers presented at the Eurotherm Conference 105: Computational Thermal Radiation in Participating Media V, which was held in Albi, France on 1-3 April 2015. This seminar was the fifth in a series after Nancy, France (Eurotherm Seminar 95, April 2012), Mons, Belgium (Eurotherm Seminar 73, April 2003), Poitiers, France (Eurotherm Seminar 78, April 2006) and Lisbon, Portugal (Eurotherm Seminar 83, April 2009).

Around 40 contributions were received during the conference preparation that have been submitted to oral presentations. A selection process based on two peer-reviews of the full papers finally resulted in the acceptance of 36 for oral presentations (including 2 plenary lectures). These 2 plenary lectures and 10 other papers have been selected for a special issue in a journal related to radiative heat transfer and will not be presented in this volume.

The conference was attended by almost 60 scientists from 15 different countries: Australia, Belgium, Canada, China, France, Germany, Poland, Portugal, Russia, Switzerland, The Netherlands, Sweden, Tunisia, Turkey and USA.

The 24 contributions presented in this volume relate to the following main topics:

- Radiative transfer in complex media
- Applications, combustion and high temperatures
- Inverse methods
- New developments for RTE solution
- Gas radiation modeling
- Particles, droplets and dispersed systems
- Monte Carlo methods
- Photoreactive processes
- Radiative transfer for medical applications
- Radiative transfer in planetary atmospheres and oceans
- Coupled heat transfer problems involving radiative effects
- Radiative transfer problems at micro - and nanoscales

The conference organizers gratefully acknowledge the members of the scientific committee and the experts who carried out the reviews of the papers, and the local organizing committee for their preparation of the conference. Also acknowledged are the support by the « investissements d'avenir and the labex Solstice », the Société Française de Thermique, ICHMT, the sponsorship of the Midi-Pyrénées Region, the Toulouse University (Université Fédérale de Toulouse), and the financial contributions of city of Albi.

Journal of Physics: Conference Series **676** (2016) 011001 doi:10.1088/1742-6596/676/1/011001

Conference Chairs

Mouna EL Hafi	Mines-Albi, France
Richard Fournier	Laplace, Université de Toulouse, France
D Lemonnier	Institut P', Poitiers, France
N Selçuk	Middle East Technical University, Ankara, Turkey
P Lybaert	Umons, Faculty of engineering, Belgium

International Scientific Committee

R Bialecki	STU Gliwice, Poland
P. Boulet	LEMTA, Nancy, France
P Coehlo	Instituto Superior Tecnico, Lisboa, Portugal
L A Dombrovsky	IHT, Moscow, Russia
J Howell	University of Texas, Austin, USA
S Kumar	Polytechnic University of New York, USA
W Lipinski	Australian National University , Australia
F Liu	Institute for Chemical Process and Environmental Technology, Canada
L.H. Liu	Harbin Inst. of Tech., China
W Malalasekera	Loughborough University, UK
S Maruyama	Institute of Fluid Science, Tohoku, Japan
M P Mengüç	Ozyeğin University, Istanbul, Turkey
M Mishchenko	NASA Goddard Institute, USA
M F Modest	University of California, USA
L Pilon	University of California, Los Angeles, USA
S Rukolaine	Ioffe Institute, St Petersburg, Russia
C Salinas	University of Campinas, Sao Paulo, Brazil
A Soufiani	École Centrale, Paris, France
R Vaillon	CETHIL INSA, Lyon, France
B Webb	Brigham Young University, USA
Z Zhang	Georgia Institute of Technology, USA
H C Zhou	Huazhong University of Science and Technology, Wuhan, China

Guest Editors

M. EL Hafi

Mines-Albi, France

mouna.elhafi@mines-albi.fr

R. Fournier

Laplace, Université de Toulouse, France

richard.fournier@laplace.univ-tlse.fr

D Lemonnier

PPrime Institute, CNRS -- ENSMA, University of Poitiers, France

denis.lemonnier@ensma.fr

P Lybaert

UMONS, Thermal Engineering and Combustion Laboratory, Belgium

paul.lybaert@umons.ac.be

N Selçuk

Middle East Technical University, Turkey

selcuk@metu.edu.tr

Journal of Physics: Conference Series **676** (2016) 011002 doi:10.1088/1742-6596/676/1/011002

Peer review statement

All papers published in this volume of *Journal of Physics: Conference Series* have been peer reviewed through processes administered by the proceedings Editors. Reviews were conducted by expert referees to the professional and scientific standards expected of a proceedings journal published by IOP Publishing.

Journal of Physics: Conference Series **676** (2016) 012001 doi:10.1088/1742-6596/676/1/012001

An exact formulation of *k*-distribution methods in non-uniform gaseous media and its approximate treatment within the Multi-Spectral framework

Frédéric ANDRE[1,3], Longfeng HOU[1], Vladimir P. SOLOVJOV[2]

[1] Université de Lyon, CNRS, INSA-Lyon, CETHIL, UMR5008, F-69621 Villeurbanne, France
[2] Brigham Young University, Department of Mechanical Engineering, CTB 435, Provo, UT 84602,USA

E-mail: frederic.andre@insa-lyon.fr

Abstract. The main restriction of *k*-distribution approaches for applications in radiative heat transfer in gaseous media arises from the use of a scaling or correlation assumption to treat non-uniform situations. It is shown that those cases can be handled exactly by using a multidimensional *k*-distribution that addresses the problem of spectral correlations without using any simplifying assumptions. Nevertheless, the approach cannot be suggested for engineering applications due to its computational cost. Accordingly, a more efficient method, based on the so-called Multi-Spectral Framework, is proposed to approximate the previous exact formulation. The model is assessed against reference LBL calculations and shown to outperform usual *k*-distribution approaches for radiative heat transfer in non-uniform media.

1. Introduction

Radiative heat transfer in gaseous media is encountered in many engineering and research studies dedicated, for instance, to the effect of radiation on the structures of planetary atmospheres or for the calculation of heat fluxes at the walls of combustion chambers. The most accurate and flexible model to treat gas radiation interactions is the so-called Line-By-Line (LBL) approach. LBL spectra are based directly on spectroscopic databases and provide gas radiative properties at high resolution. Nevertheless, this method is computationally too expensive to be used in many situations. For most engineering applications, simpler models are required.

Many approximate methods have been proposed during the past decades to treat radiative heat transfer in gaseous media. A compendium of the existing ones can be found for instance in Refs. [1,2]. Among them, *k*-distribution techniques are currently the most widely used, partly due to their ability to be applied together with any Radiative Transfer Equation (RTE) solver. They are usually accurate and can be readily extended to any band width, from narrow bands up to the Full Spectrum (FS), with minor modifications. The so-called ADF [3], FSK [4] and SLW [5] models are undoubtedly the most famous FS *k*-distribution approaches and have been applied to many engineering configurations. Over narrow bands, the C*k* (Correlated-*k*) [1] model is recognized as one of the most relevant choices to treat gas radiation. All those models are essentially based on the same set of two equations (the first one for the *k*-distribution formulation in uniform media and the second one for its extension to non-

[3] To whom any correspondence should be addressed.

uniform cases – this second equation is usually based on a scaling or correlated assumption of gas spectra in distinct thermophysical states). Those models provide good accuracy for most engineering applications but may fail to give reliable results when high temperature gradients are encountered along the radiation path. This is mainly due to the appearance of "hot lines" (at high temperature) that are missing in the gas spectra at low temperatures. The effect of these lines is to break the "ideal" behaviour of the gas spectra and may reduce drastically the accuracy of the approximate models based on this assumption. This problem (encountered for instance in IR plume signature, or in some emission spectroscopy configurations) can be solved by application of the so-called fictitious gases approach [6] or by using mapping techniques [7,8]. Those methods improve the accuracy of the approximate model but usually at the cost of a large increase in terms of computational resources.

The aim of the present paper is to show that *k*-distributions can be formulated exactly in non-uniform media by replacing the usual *k*-distribution concept (that is defined for a given thermophysical state) by a multidimensional one. Using this exact model in engineering applications cannot be suggested in a general frame but approximate solutions can be proposed. The method to approximate this exact solution for radiative transfer applications was described in a previous paper and led us to develop the so-called "Multi-Spectral Framework - MSF" [8]. It will be shown that all existing models are special cases of the MSF. This point is raised in Section 4. Results obtained through the application of this technique are assessed against reference LBL calculations in situations representative of engineering problems, in 0D (along line-of-sights) and 1D (plane parallel walls) geometries.

2. *k*-distribution methods

k-distribution approaches were introduced in 1934 by Ambartzumian [9]. At this time, due to a lack of information about the real profiles of spectral lines, the absorption coefficient was assumed to be a periodic function of the wavelength. The same technique was used to treat real gas spectra (without any name but with reference to Ambartzumian's work) during the following decades (see for instance Refs. [10,11]). The name "*k*-distribution" only appeared (to our knowledge) in 1972 [12]. Since then, many papers have been devoted to its application both in uniform (homogeneous isothermal) and non-uniform media [1,2].

In these methods, the (narrow) band averaged transmission function of a path of length L inside a gas at a uniform temperature T, pressure P, and molar fraction in absorbing specie x is expressed as:

$$\tau^{\Delta\eta}(L)=\frac{1}{\Delta\eta}\int_{\Delta\eta}\exp\left(-xP\kappa_\eta L\right)\,d\eta = \int_0^{+\infty}\exp\left(-xPLk\right)f\left(k\right)dk \tag{1}$$

where $f(k)$ is the *k*-distribution function. For any real interval $\left[k,k+dk\right]$, $dk>0$, $f(k)\,dk$ represents the fraction of wavenumbers $\eta\in\Delta\eta$ such that the value of the spectral absorption coefficient of the gas is inside the interval $\left[k,k+dk\right]$. Eq. (1) can also be written in terms of the cumulative distribution of *k*-values, $g(k)$, as:

$$\tau^{\Delta\eta}(L)=\int_0^{+\infty}exp\left(-xPLk\right)dg\left(k\right) \tag{2}$$

Eq. (2) mainly consists of a change of variable of integration $\eta\leftrightarrow k$ (or equivalently $\eta\leftrightarrow g(k)$ as g is strictly monotonous). Therefore, it provides the exact band averaged transmission function. The main source of error arises in practice from the use of numerical quadratures to estimate this integral. Mathematically, $g(k)$ associates to any value of $k\in\mathbb{R}$ a probability to find a value of κ_η lower than k inside $\Delta\eta$. This probability $\mathbb{P}$ is defined as the fraction of the interval $\Delta\eta$ such that $\kappa_\eta<k$. It can be written formally as [1]:

$$\mathbb{P}\left(\kappa_\eta < k, \eta \in \Delta\eta\right) = g(k) = \frac{1}{\Delta\eta} \int_{\Delta\eta} H\left(k-\kappa_\eta\right) d\eta \Leftrightarrow f(k) = \frac{dg(k)}{dk} = \frac{1}{\Delta\eta} \int_{\Delta\eta} \delta\left(k-\kappa_\eta\right) d\eta \quad (3)$$

In Eq. (3), H is the Heaviside function and δ is the Dirac delta function defined as [13] (where F is any continuous function defined over the real axis and c is a constant):

$$\delta(x-c)=0,\ x\neq c \quad (4)$$

$$\int_{-\infty}^{+\infty} \delta(x-c)\, F(x)\, dx = F(c) \quad (5)$$

The notation used in the second term at the RHS in Eq. (3) is purely formal. In fact, "function" $\delta\left(k-\kappa_\eta\right)$ is mathematically defined only if the absorption coefficient is strictly monotonous over the interval (viz. if the derivative of the absorption coefficient with respect to the wavenumber does not vanish, [13]) which is not the case in a general frame. Furthermore, this definition of the distribution function only has a sense inside an integral over k (such as in Eq. (1)). Accordingly, in the present paper, as soon as possible, equations will make use of g, which defines a measure on the spaces of k-values and that can be written in terms of Heaviside function as shown in Eq. (3), rather than f. Nevertheless, as this notation is usual in gas radiation, we will make use of $\delta\left(k-\kappa_\eta\right)$ in the paper. Consequently, all spectral bands must be understood as the restrictions of the intervals over which $\delta\left(k-\kappa_\eta\right)$ is mathematically defined. As the number of minima and maxima of the absorption coefficient are always finite inside a spectral interval, the values of the spectral integrals are not changed if we consider the full band or those restricted sets of wavenumbers.

k-distribution approaches can be extended to non-uniform media by assuming "ideal" behaviours of the variations of the absorption coefficient with respect to the thermophysical conditions. The most widely used assumptions are called the scaling and correlation approximations [14]. They are briefly described in the next section.

3. Multidimensional *k*-distribution methods in non-uniform media

3.1. *Preliminary comments*

In this section, we introduce the method proposed to extend k-distribution approaches used in uniform media to non-uniform ones. The technique is detailed in a simple case where the path is discretized in two uniform sub-paths in distinct thermophysical states. The approach is then extended to the general case where the non-uniform path can be divided into $n \in \mathbb{N}$ homogeneous isothermal elements. Furthermore, the only kind of path non-uniformity considered here will be due to temperature. This is not a limitation of the method, as it can also be applied when gradients of composition or pressure are encountered along the path, but this assumption is made to simplify the notations.

Only spectral narrow bands will be considered in this work.

3.2. *Exact treatment of spectral correlation effects*

Let us consider a narrow spectral interval $\Delta\eta$ that: 1/ is narrow enough to assume that the Planck function remains constant, 2/ do not contain any transparency region of the gas at any temperature (this means that $\forall\eta\in\Delta\eta,\ \forall T\in\left[T_{\min}, T_{\max}\right],\ \kappa_\eta(T)>0$ where $T_{\min}$ and $T_{\max}$ represent the minimum and maximum temperatures encountered along the non-uniform path).

Over band $\Delta\eta$, it is possible to exhibit relationship between absorption coefficients $\kappa_\eta(T_1)$ and $\kappa_\eta(T_2)$ at two distinct temperatures, $T_1 \neq T_2$ arbitrarily chosen inside the interval $\left[T_{\min}, T_{\max}\right]$, by plotting one of them (for instance $\kappa_\eta(T_2)$) as a function of the other one (in this case $\kappa_\eta(T_1)$). This

Journal of Physics: Conference Series **676** (2016) 012001 doi:10.1088/1742-6596/676/1/012001

approach provides a 2-dimensional parametric curve $k_1 = \kappa_\eta(T_1)$, $k_2 = \kappa_\eta(T_2)$ in the (k_1, k_2)-plane, that will be subsequently referred to as the C curve. This curve can be interpreted in the following way:

- If C is a straight line $k_2 = U \cdot k_1$ with $U > 0$, then for any wavenumber $\eta \in \Delta\eta$ we have $\kappa_\eta(T_2) = U \cdot \kappa_\eta(T_1)$, and then the two absorption coefficients are called scaled.
- If C is a function $k_2 = F(k_1)$, then for any value of $\eta \in \Delta\eta$, we have $\kappa_\eta(T_2) = F\left[\kappa_\eta(T_1)\right]$ and the two absorption coefficients are correlated. It should be noticed that in the literature on gas radiation, the concept of correlation usually involved is more restrictive since it assumes F to be strictly monotonous [1].

In other cases, no conclusion about correlation can be made: generally, when absorption coefficients associated to two distinct temperatures are plotted as a function of each other, none of the two previous situations (viz. scaling or correlation) is rigorously encountered.

To illustrate this fact, we consider the two spectra plotted in figure 1. They correspond to two absorption coefficients given by the Elsasser model (infinite array of regularly spaced identical Lorentz lines). As those absorption coefficients are assumed to be associated to distinct thermophysical states, spectra in figure 1 were obtained by using distinct values of the Half Widths at Half Maximum (HWHM) of the Lorentz profiles (to account for the effect of pressure, composition and temperature changes between the two states) as well as of the line intensity (known to be strongly temperature dependent). Furthermore, the distance between lines was chosen distinct in the two spectra to account for the possible appearance of "hot" lines. The corresponding parametric curve C, based on the previous absorption coefficients, is shown in figure 2.

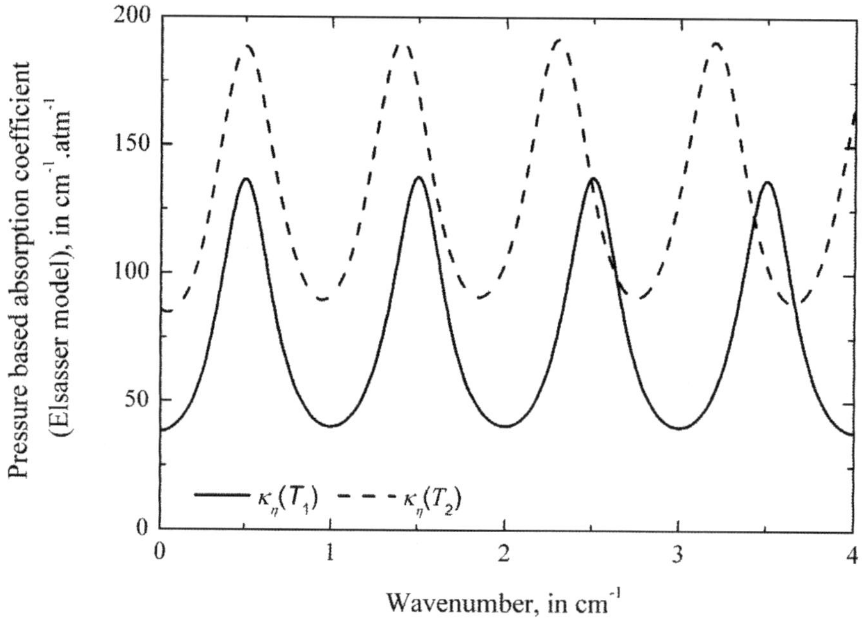

Figure 1. Elsasser model of absorption coefficients at two distinct thermophysical states.

Journal of Physics: Conference Series **676** (2016) 012001 doi:10.1088/1742-6596/676/1/012001

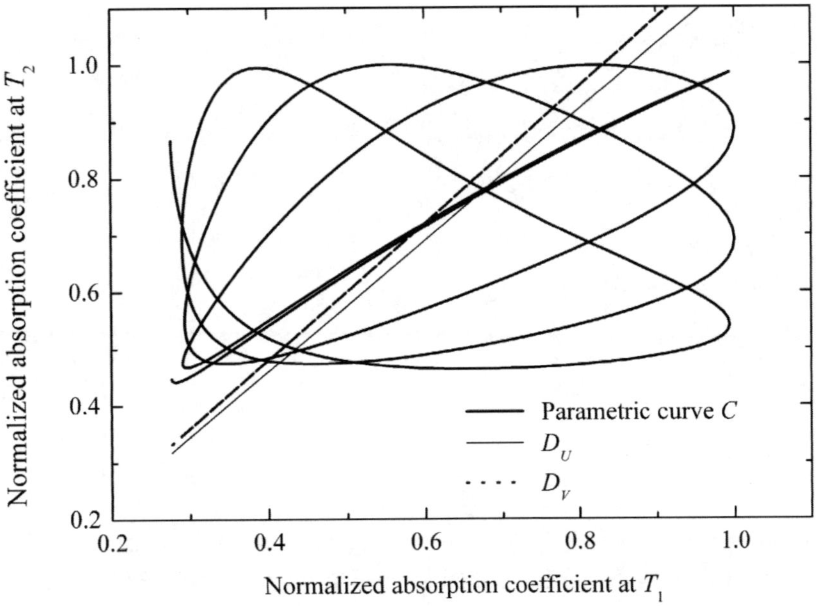

Figure 2. Parametric curve C for the two absorption spectra depicted in figure 1.

Our objective is to calculate the mean value of some continuous real valued function F of both $\kappa_\eta(T_1)$ and $\kappa_\eta(T_2)$ averaged over band $\Delta\eta$:

$$F^{\Delta\eta} = \frac{1}{\Delta\eta}\int_{\Delta\eta} F\left[\kappa_\eta(T_1),\kappa_\eta(T_2)\right] d\eta \tag{6}$$

As $\kappa_\eta(T_1)$ and $\kappa_\eta(T_2)$ are neither scaled nor correlated, it is not easy to suggest a simple method to handle the calculation of this integral. One possible technique is described below.

Let us introduce two straight lines, that will be called D_U and D_V, whose equations are given respectively as $k_2 = U\cdot k_1$ and $k_2 = V\cdot k_1$ with $V > U > 0$ (see figure 2). Any point on curve C with coordinates $(k_1 = \kappa_\eta(T_1),\ k_2 = \kappa_\eta(T_2))$ located between D_U and D_V is such that:

$$U < \kappa_\eta(T_2)/\kappa_\eta(T_1) < V \tag{7}$$

or equivalently:

$$U < u_\eta(T_1,T_2) < V \tag{8}$$

where we define the function $u_\eta(T_1,T_2) = \kappa_\eta(T_2)/\kappa_\eta(T_1)$ to represent the spectral scaling function between the two absorption spectra.

The fraction of the interval $\Delta\eta$ for which Eqs. (7,8) are satisfied is:

$$\frac{\Delta\eta(U,V)}{\Delta\eta} = \frac{1}{\Delta\eta}\int_U^V \left\{\int_{\Delta\eta} \delta\left[u - u_\eta(T_1,T_2)\right] d\eta\right\} du \tag{9}$$

and the mean value of function F over the same set of wavenumbers is given as:

$$F^{\Delta\eta(U,V)} = \int_U^V \left\{ \frac{1}{\Delta\eta(U,V)} \int_{\Delta\eta} \delta\left[u - u_\eta(T_1,T_2)\right] \; F\left[\kappa_\eta(T_1),\kappa_\eta(T_2)\right] d\eta \right\} du \tag{10}$$

Since variables η and u are independent, the following (purely formal) equality:

$$\int_{\Delta\eta} \delta\left[u - u_\eta(T_1,T_2)\right] \; F\left[\kappa_\eta(T_1),\kappa_\eta(T_1)\cdot u_\eta(T_1,T_2)\right] d\eta = \int_{\Delta\eta} \delta\left[u - u_\eta(T_1,T_2)\right] \; F\left[\kappa_\eta(T_1),u\cdot\kappa_\eta(T_1)\right] d\eta \tag{11}$$

enables to write Eq. (10) as:

$$F^{\Delta\eta(U,V)} = \frac{1}{\Delta\eta(U,V)} \int_{\Delta\eta} \left\{ \int_U^V \delta\left[u - u_\eta(T_1,T_2)\right] \; F\left[\kappa_\eta(T_1),u\cdot\kappa_\eta(T_1)\right] du \right\} d\eta \tag{12}$$

Now, we observe that:

$$\lim_{\substack{U\to 0\\ V\to+\infty}} \Delta\eta(U,V) = \Delta\eta \tag{13}$$

Eq. (13) simply means that when $U \to 0$, then D_U approaches the k_1-axis, and if $V \to +\infty$, then D_V approaches the k_2-axis. In this case, the domain between the two lines covers the first quarter of the real plane, to which any value of $\kappa_\eta(T_1)$ and $\kappa_\eta(T_2)$ belong.
Using Eq. (13), we can then conclude that:

$$\lim_{\substack{U\to 0\\ V\to+\infty}} F^{\Delta\eta(U,V)} = F^{\Delta\eta} = \frac{1}{\Delta\eta} \int_{\Delta\eta} \left\{ \int_0^{+\infty} \delta\left[u - u_\eta(T_1,T_2)\right] \; F\left[\kappa_\eta(T_1),u\cdot\kappa_\eta(T_1)\right] du \right\} d\eta \tag{14}$$

The previous equation can also be written by using the property of the Delta function given by Eq. (5) that provides $F\left[\kappa_\eta(T_1),u\cdot\kappa_\eta(T_1)\right] = \int_0^{+\infty} \delta\left[k - \kappa_\eta(T_1)\right] F(k,u\cdot k)\, dk$:

$$F^{\Delta\eta} = \int_0^{+\infty}\int_0^{+\infty} \left\{ \frac{1}{\Delta\eta} \int_{\Delta\eta} \delta\left[u - u_\eta(T_1,T_2)\right] \; \delta\left[k - \kappa_\eta(T_1)\right] d\eta \right\} F(k,u\cdot k)\, dk\, du \tag{15}$$

Eq. (15) can then be reformulated using the change of variables $k_1 = k,\ k_2 = u\cdot k$. Indeed, the Jacobian corresponding to this change of variable is k (which is strictly positive, as assumed in section 3.2) and, using Eq. (21) page 53 from Ref. [13], we can write formally here: $\delta\left[u - u_\eta(T_1,T_2)\right]\cdot\delta\left[k - \kappa_\eta(T_1)\right] = k\cdot\delta\left[k_1 - \kappa_\eta(T_1)\right]\cdot\delta\left[k_2 - \kappa_\eta(T_2)\right]$. This provides

$$F^{\Delta\eta} = \int_0^{+\infty}\int_0^{+\infty} \left\{ \frac{1}{\Delta\eta} \int_{\Delta\eta} \delta\left[k_1 - \kappa_\eta(T_1)\right] \; \delta\left[k_2 - \kappa_\eta(T_2)\right] d\eta \right\} F(k_1,k_2)\, dk_1\, dk_2 \tag{16}$$

This equation generalizes the usual k-distribution approach given by Eq. (1) to non-uniform media. Indeed, the previous relation can be written as:

$$F^{\Delta\eta} = \int_0^{+\infty}\int_0^{+\infty} F(k_1,k_2)\, dg(k_1,k_2) \tag{17}$$

in which we have introduced the 2-dimensional cumulated k-distribution:

$$g(k_1,k_2)=\frac{1}{\Delta\eta}\int_{\Delta\eta} H\left[k_1-\kappa_\eta(T_1)\right] H\left[k_2-\kappa_\eta(T_2)\right] d\eta \tag{18}$$

It should be noticed that, as in the case of *k*-distribution methods for uniform media, using distribution $g(k_1,k_2)$ provides an exact value of the spectral average of $F\left[\kappa_\eta(T_1),\kappa_\eta(T_2)\right]$ over $\Delta\eta$. Now, let us consider the particular case:

$$F\left[\kappa_\eta(T_1),\kappa_\eta(T_2)\right]=exp\left[-x_1P_1\ \kappa_\eta(T_1)L_1-x_2P_2\ \kappa_\eta(T_2)L_2\right] \tag{19}$$

where L_1 and L_2 are the lengths of two uniform columns at temperature T_1 and T_2, compositions x_i, $i=1,2$ and total pressures P_i, $i=1,2$ respectively. $F^{\Delta\eta}$ represents in this case the band averaged transmission function $\tau^{\Delta\eta}(L)$ of a non-uniform path $L=L_1+L_2$. Application of Eq. (17) to this function enables to calculate exactly the transmissivity of the non-uniform path.

Furthermore, Eq. (17) remains exact not only for this non-uniform case but also for uniform situations (at temperature T_1 or T_2 that are particular cases of Eq. (17) with L_1 = 0 or L_2 = 0). In fact, in these situations, Eq. (17) simplifies to the usual *k*-distribution model. Additionally, it is worth noticing that the exactness of the approach does not rely on assumptions of scaling or correlation about gas spectra, which are known to be the main source of error for the application of *k*-distribution techniques in non-uniform media.

The simple process that we have described in this section for the treatment of a non-uniform gas path at two distinct temperatures can be readily applied to real gas spectra (instead of the Elsasser model, as considered previously). An example of the parametric curve *C* in a real case is given in figure 3. It corresponds to two spectra of H_2O over the [3287.5 ; 3292.5] cm^{-1} spectral interval for T_1 = 300 K and T_2 = 2,300 K – additional details about the LBL data used for the curve are given in the caption to figure 3.

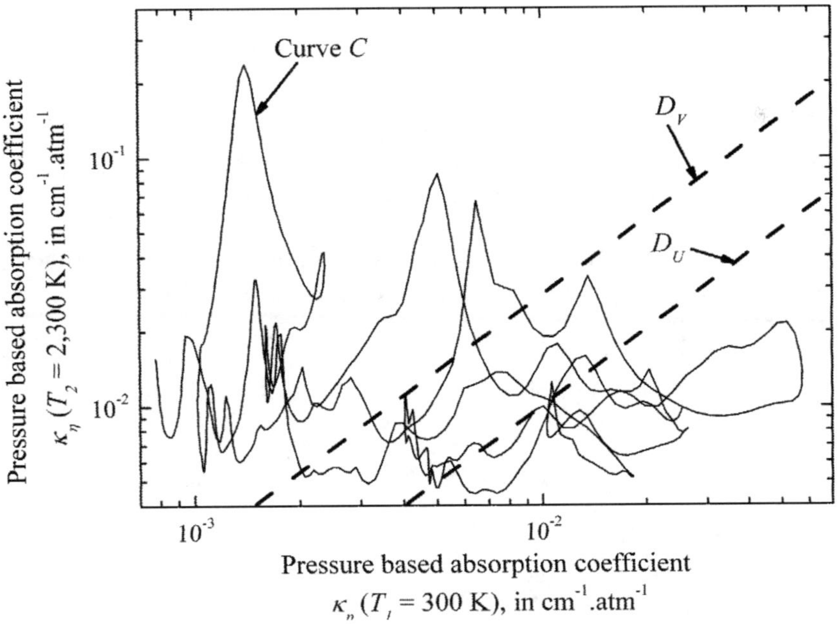

Figure 3. Parametric curve *C* representing $\kappa_\eta(T_2)$ as a function of $\kappa_\eta(T_1)$ for H_2O at T_1 = 300 K and T_2 = 2,300 K – 10% of H_2O with 90% N_2 – values of κ_η are considered over the [3287.5 ; 3292.5] cm^{-1} spectral interval.

Clearly, since curve C is not monotonous, the two spectra are neither scaled nor correlated. Notice that in log-log scales, as used on this figure, lines D_U and D_V are parallel with a slope of 1. The constant distance between D_U and D_V is $ln(V/U)$.

The approach can also be generalized to calculate band averaged values of functions F of any set of distinct temperatures $\{T_1, T_2, .., T_n\}$ where n is the number of isothermal sub-layers encountered along the non-uniform path. The method is then fundamentally the same as the one given in this section (but then parametric curves C in dimensions higher than 2 have to be considered). This provides, as a generalization of Eqs. (17,18):

$$\begin{aligned} F^{\Delta\eta} &= \frac{1}{\Delta\eta} \int_{\Delta\eta} F\left[\kappa_\eta(T_1), \kappa_\eta(T_2), .., \kappa_\eta(T_n)\right] d\eta \\ &= \underbrace{\int_0^{+\infty} \int_0^{+\infty} .. \int_0^{+\infty}}_{n \text{ integrals}} F(k_1, k_2, .., k_n)\, dg_n(k_1, k_2, .., k_n) \end{aligned} \tag{20}$$

where the n-dimensional cumulated k-distribution is defined as:

$$g_n(k_1, k_2, .., k_n) = \frac{1}{\Delta\eta} \int_{\Delta\eta} H\left[k_1 - \kappa_\eta(T_1)\right] H\left[k_2 - \kappa_\eta(T_2)\right] ... H\left[k_n - \kappa_\eta(T_n)\right] d\eta \tag{21}$$

Eq. (20) can also be written in a simpler form by introducing the vector $\boldsymbol{k} = (k_1,\ k_2,\ ..,\ k_n) \in \mathbb{R}_+^n$:

$$F^{\Delta\eta} = \int_{\boldsymbol{k} \in \mathbb{R}_+^n} F(\boldsymbol{k})\, dg_n(\boldsymbol{k}) \tag{22}$$

together with the following notations:

$$g_n(\boldsymbol{k}) = g_n(k_1, k_2, .., k_n) \tag{23-a}$$

$$F(\boldsymbol{k}) = F(k_1, k_2, .., k_n) \tag{23-b}$$

One of the main advantages of the last formulation is that it is formally the same as encountered in usual k-distribution approaches. Eq. (22) may also be extended from discrete to continuous temperature distributions along a non-uniform path using a functional integral form as shown in Appendix.

Finally, the presented method is not restricted to non-uniformities in terms of temperature but can also be applied in cases of gradients of pressure or species concentrations. This enables to write formally Eq. (21) in a more general form:

$$g_n(k_1, k_2, .., k_n) = \frac{1}{\Delta\eta} \int_{\Delta\eta} H\left[k_1 - \kappa_\eta(\underline{\phi_1})\right] H\left[k_2 - \kappa_\eta(\underline{\phi_2})\right] ... H\left[k_n - \kappa_\eta(\underline{\phi_n})\right] d\eta \tag{24}$$

where $\underline{\phi_i}$, $i = 1, .., n$ represent the state vectors whose components are the gas temperatures, total pressures, and species concentration encountered along the non-uniform path.

4. The Multi-Spectral Framework (MSF).

4.1. Principle of the MSF

One of the main problems for the application of Eq. (22) for radiative heat transfer is that it requires to estimate the multidimensional cumulative k-distribution $g_n(\boldsymbol{k}) = g(k_1, k_2, .., k_n)$, which can be computationally expensive if n is large. However, this approach can be considered in some cases (see for instance Ref. [7], in which a similar process is applied in the context of radiative transfer in the atmosphere). Nevertheless, a more efficient method is required for most engineering applications. One of them is the so-called Multi-Spectral Framework (MSF) introduced in Ref. [8]. It is out of the scope of the present work to provide a comprehensive description of this approach, and more detail can be found in Ref. [8].

In the MSF, the first step is to consider a particular case of Eq. (22) in which temperatures $\{T_1,..,T_n\}$ are chosen distinct and regularly spaced between $T_{\min}$ and $T_{\max}$: $T_{\min}=T_1<T_2<...<T_n=T_{\max}$. In the second step, it is assumed that it is possible to define P distinct and linearly independent continuous scaling functions $u_p(T)$, $p=1,..,P$ such that for any wavenumber $\eta\in\Delta\eta$, one can find an integer $p\in\{1,..,P\}$ such that:

$$\forall i\in\{1,..,n\},\ \kappa_\eta(T_i)=\kappa_\eta(T^{ref})\ u_p(T_i) \tag{25}$$

where T^{ref} represents a reference temperature, chosen arbitrarily inside the interval $[T_{\min},T_{\max}]$. Here, we choose the reference temperature $T^{ref}=T_1$, that implies: $u_p(T_1)=1$. Eq. (25) allows definition of P spectral subintervals $\Delta\eta(u_p)$, $p=1,..,P$: $\bigcup_{p=1,P}\Delta\eta(u_p)=\Delta\eta$ that can be used to discretize Eq. (22) into:

$$F^{\Delta\eta}=\frac{1}{\Delta\eta}\sum_{p=1}^{P}\Delta\eta(u_p)\cdot\int_{\boldsymbol{k}\in\mathbb{R}_+^n}F(\boldsymbol{k})\ dg_{n,p}(\boldsymbol{k}) \tag{26}$$

where, as a direct application of Eq. (21):

$$g_{n,p}(\boldsymbol{k})=\frac{1}{\Delta\eta(u_p)}\int_{\Delta\eta(u_p)}H\left[k_1-\kappa_\eta(T_1)\right]H\left[k_2-\kappa_\eta(T_2)\right]...H\left[k_n-\kappa_\eta(T_n)\right]d\eta \tag{27}$$

and where $\Delta\eta(u_p)$ is the width of the spectral interval set by Eq. (25).

It can be noticed that it is always possible to define such a finite set of scaling functions because in practice LBL data are calculated at limited resolution (which means that inside an interval, the number N of spectral values of the absorption coefficient is finite and the number of distinct scaling functions is utmost equal to N). Then, the third step consists in using the definition of the scaling functions given by Eq. (25) inside Eq. (27) to set:

$$g_{n,p}(\boldsymbol{k})=\frac{1}{\Delta\eta(u_p)}\int_{\Delta\eta(u_p)}H\left[k_1-\kappa_\eta(T^{ref})\right]H\left[k_2-\kappa_\eta(T^{ref})\cdot u_p(T_2)\right]...H\left[k_n-\kappa_\eta(T^{ref})\cdot u_p(T_n)\right]d\eta \tag{28}$$

It can be shown (see Eq. (27) in Ref. [8]) that Eq. (28) simplifies into:

$$g_{n,p}(\boldsymbol{k})=\frac{1}{\Delta\eta(u_p)}\int_{\Delta\eta(u_p)}H\left[k_1-\kappa_\eta(T^{ref})\right]d\eta=g_p(k_1,T^{ref}) \tag{29}$$

where $g_p(k_1,T^{ref})$ is the distribution of k-values at the reference temperature $T^{ref}=T_1$ inside the interval $\Delta\eta(u_p)$.

Accordingly, Eq. (26) can be written in a very simple form (see Eq. (26) in Ref. [8]):

$$F^{\Delta\eta}=\frac{1}{\Delta\eta}\sum_{p=1}^{P}\Delta\eta(u_p)\int_0^{+\infty}F\left[k_1,k_1\cdot u_p(T_2),...,k_1\cdot u_p(T_n)\right]dg_p(k_1,T^{ref}) \tag{30}$$

The application of the MSF thus reduces the calculation of a multidimensional k-distribution to a weighted sum of usual ones. Furthermore, for any spectrum of the gas at a temperature $T \in [T_{\min}, T_{\max}]$, it is possible to approximate $\kappa_\eta(T)$ over any spectral interval $\Delta\eta(u_p)$, $p = 1,..,P$ as:

$$\kappa_\eta(T) \approx \kappa_\eta(T_1) \cdot u_p(T),\ \eta \in \Delta\eta(u_p) \tag{31}$$

This provides, for instance for the transmission function of the gas at temperature T :

$$\tau^{\Delta\eta}(L) = \frac{1}{\Delta\eta} \int_{\Delta\eta} exp\left[-xP\kappa_\eta(T)L\right] d\eta \approx \frac{1}{\Delta\eta} \sum_{p=1}^{P} \Delta\eta(u_p) \int_0^{+\infty} exp\left[-xP\, k_1 \cdot u_p(T)\, L\right] dg_p\left(k_1, T^{ref}\right) \tag{32}$$

Obviously, this approach can also be extended to non-uniform paths.
The main difficulty for the application of the previous equations is the definition of the set of scaling functions $u_p(T)$, $p = 1,..,P$. This point is raised in the next section.

4.2. Application of a functional clustering technique to define $\Delta\eta(u_p)$, $p = 1,..,P$

Let us consider an arbitrary integer $p \in \{1,..,P\}$ and two arbitrary distinct wavenumbers $\eta_1, \eta_2 \in \Delta\eta(u_p)$. Following the definition of intervals $\Delta\eta(u_p)$ (see Eq. (25)), we have:

$$C(\eta_1,\eta_2) = \int_{T_{\min}}^{T_{\max}} \kappa_{\eta_1}(T)\kappa_{\eta_2}(T)dT = \kappa_{\eta_1}(T^{ref})\kappa_{\eta_2}(T^{ref}) \int_{T_{\min}}^{T_{\max}} \left[u_p(T)\right]^2 dT \tag{33}$$

and:

$$C(\eta_1,\eta_1)C(\eta_2,\eta_2) = \int_{T_{\min}}^{T_{\max}} \left[\kappa_{\eta_1}(T)\right]^2 dT \int_{T_{\min}}^{T_{\max}} \left[\kappa_{\eta_2}(T)\right]^2 dT = \left[\kappa_{\eta_1}(T^{ref})\kappa_{\eta_2}(T^{ref})\right]^2 \left\{\int_{T_{\min}}^{T_{\max}} \left[u_p(T)\right]^2 dT\right\}^2 \tag{34}$$

that shows that:

$$\frac{C(\eta_1,\eta_2)}{\left[C(\eta_1,\eta_1)C(\eta_2,\eta_2)\right]^{1/2}} = 1 \tag{35}$$

Similarly, if the two wavenumbers are chosen in distinct spectral intervals $\Delta\eta(u_p)$ and $\Delta\eta(u_{p'})$, $p \neq p'$ then, according to the Cauchy-Schwartz inequality [15]:

$$0 \leq \frac{C(\eta_1,\eta_2)}{\left[C(\eta_1,\eta_1)C(\eta_2,\eta_2)\right]^{1/2}} = \frac{\int_{T_{\min}}^{T_{\max}} u_p(T)u_{p'}(T)dT}{\left\{\int_{T_{\min}}^{T_{\max}} \left[u_p(T)\right]^2 dT\right\}^{1/2} \left\{\int_{T_{\min}}^{T_{\max}} \left[u_{p'}(T)\right]^2 dT\right\}^{1/2}} < 1 \tag{36}$$

Consequently, the two criteria set by Eqs. (35,36) can be used to group together the wavenumbers that have the same scaling functions and thus to determine the spectral intervals $\Delta\eta(u_p)$. This process is usually referred to as Functional Data Clustering [16], which is a part of more general Functional Data Analysis (FDA) [17].

FDA is a relatively new area of mathematical science whose aim is to extend the usual techniques encountered for Multivariate Statistical Analysis (MSA) of data (such as clustering techniques for instance) to functions. The steps for Functional Data Clustering are [18]:

1/ Propose a functional form for the data.

2/ Estimate the parameters that provide the best estimate of the data using the functional form chosen in the previous step.

3/ Calculate a similarity (see Eq. (35)) or a distance between functions based on integrals (and not in terms of distances between points, which is the most significant difference between FDA and MSA).

4/ Apply usual techniques encountered in MSA (such as clustering, principal component analysis, etc) using the coefficients of similarities (or distances) obtained at the previous stage to build groups (here spectral intervals) associated to similar functions.

Table 1 summarizes the two functional forms of the scaling function (step 1) considered in this work. The first one was used previously in Ref. [8]. It can be shown that the two formulas given in the third column converge toward their exact value (see Eq. (37)) when $T_i \to T_{i+1}$.

Table 1. Functional forms used for the definition of the similarity coefficient required in the clustering process and associated cross-correlation coefficients over $[T_i, T_{i+1}]$, $T_i \in \{T_1,..,T_n\}$.

Functional Form	Approximation of $\ln[u_\eta(T)/u_\eta(T_i)]$ over an interval $[T_i, T_{i+1}]$, α is a coefficient	$C_{i,i+1}(\eta_1,\eta_2)$ (see Eq. (37)) Spectral scaling functions $u_{\eta_j}(T)$, $j = 1,2$ are defined here as: $u_{\eta_j}(T)=\kappa_{\eta_j}(T)\left\{\int_{T_{min}}^{T_{max}}\left[\kappa_{\eta_j}(T')\right]^2 dT'\right\}^{-1/2}$
FF1	$\alpha(T-T_i)$	$\dfrac{u_{\eta_1}(T_{i+1})u_{\eta_2}(T_{i+1})-u_{\eta_1}(T_i)u_{\eta_2}(T_i)}{\ln\left[u_{\eta_1}(T_{i+1})u_{\eta_2}(T_{i+1})\right]-\ln\left[u_{\eta_1}(T_i)u_{\eta_2}(T_i)\right]}(T_{i+1}-T_i)$
FF2	$\alpha\ln(T/T_i)$	$\dfrac{u_{\eta_1}(T_i)u_{\eta_2}(T_i)T_i}{\alpha_{\eta_1}+\alpha_{\eta_2}+1}\left[\left(\dfrac{T_{i+1}}{T_i}\right)^{\alpha_{\eta_1}+\alpha_{\eta_2}+1}-1\right]$, $\alpha_{\eta_j}=\dfrac{\ln\left[u_{\eta_j}(T_{i+1})/u_{\eta_j}(T_i)\right]}{\ln(T_{i+1}/T_i)}$

The third column in table 1 enables to calculate the correlation coefficients that appear in Eqs. (35,36) as (for instance for $C(\eta_1,\eta_2)$, and omitting the product $\kappa_{\eta_1}(T^{ref})\kappa_{\eta_2}(T^{ref})$ for simplicity):

$$C(\eta_1,\eta_2)=\int_{T_{min}}^{T_{max}} u_{\eta_1}(T)u_{\eta_2}(T)dT=\sum_{i=1}^{n-1}\left[\int_{T_i}^{T_{i+1}} u_{\eta_1}(T)u_{\eta_2}(T)dT\right]=\sum_{i=1}^{n-1}C_{i,i+1}(\eta_1,\eta_2) \quad (37)$$

Those coefficients can be used in Eqs. (35,36) to evaluate the similarity (in terms of scaling functions) between any couple of wavenumbers. The wavenumbers associated to the highest values of this coefficient can then be grouped into clusters (here we have used a hierarchical clustering based on an average linkage scheme [16] – this method was shown in Ref. [19] to be well suited for the clustering of functional data). Once clusters are built, we have groups of wavenumbers that share similar scaling functions. The width $\Delta\eta(u_p)$ of the corresponding spectral interval can be estimated as $\Delta\eta(u_p)=N_p\,\delta\eta$ where N_p represents the number of elements in the p-th cluster (the one associated to the p-th scaling function $u_p(T)$) and $\delta\eta$ is the distance (assumed constant over $\Delta\eta$) between two successive values of the absorption coefficients in the LBL dataset.

4.3. Approximate treatment of non-uniformities over clusters

Following the method described in the previous subsection to build spectral intervals $\Delta\eta(u_p)$, $p=1,..,P$, the most natural way to evaluate the radiative properties of the gas over each cluster of wavenumbers is by using a Scaled-*k* approach. The main problem is that over those clusters, absorption coefficients are not rigorously scaled, but only share similar behaviors (in other words, the application of a clustering technique only ensures that the elements that are inside a same cluster are more similar with respect to the coefficients given by Eqs. (35,36) than those that are outside this cluster). This is shown in figure 4. As a consequence, functions $\hat{u}_p(\underline{\phi})=u_p(T)\cdot u_p^{'}(x,P)$, where $\hat{u}_p(\underline{\phi})$ represents the scaling function between states $\underline{\phi}^{ref}$ and $\underline{\phi}$ and $u_p^{'}(x,P)$ is the part of the function related to composition and pressure, cannot be defined explicitly but only implicitly as a solution of the following implicit equation [14] (over each interval $\Delta\eta(u_p)$):

$$\int_0^{+\infty} exp(-xPkL)\; dg_p(k,\underline{\phi})=\int_0^{+\infty} exp\left[-x^{ref}P^{ref}k^{ref}L\,\hat{u}_p(\underline{\phi})\right]\, dg_p\left(k^{ref},\underline{\phi}^{ref}\right) \qquad (38)$$

where $g_p(k,\underline{\phi})$ is the cumulative distribution of values of k in the thermophysical state $\underline{\phi}=\{\,x,\,P,\,T\,\}$ over the interval $\Delta\eta(u_p)$ and $\underline{\phi}^{ref}$ represents a prescribed reference thermophysical state. It can be shown easily that the previous equation does not provide a solution that only changes with the thermophysical state of the gas but that it also depends on the length of the path considered in Eq. (38). This means that this equation needs to be solved iteratively along the propagation of the radiation inside the gaseous medium, which is not convenient for applications in radiative heat transfer.

In order to circumvent this difficulty, we notice that as over clusters absorption coefficients are mostly linearly correlated, the use of the correlation assumption is here justified. Accordingly, instead of equation (38), we define, for convenience, the absorption coefficient $k(\underline{\phi})$ in any state $\underline{\phi}$ as a solution of the well known correlation equation [1], whose application is fully justified here because the correlating function is strictly increasing (since $\hat{u}_p(\underline{\phi})>0$):

$$g_p(k,\underline{\phi})=g_p\left(k^{ref},\underline{\phi}^{ref}\right) \qquad (39)$$

Using this method, the local absorption coefficients can be calculated once and then stored in tables. Other practical advantages of this treatment are that it allows: 1/ the use of the approach in all existing codes based on the Correlated-*k* approach, with only minor modifications, 2/ the building of the clusters for a single reference composition and pressure. The correlation assumption then deals with variations of composition and pressure along heterogeneous paths.

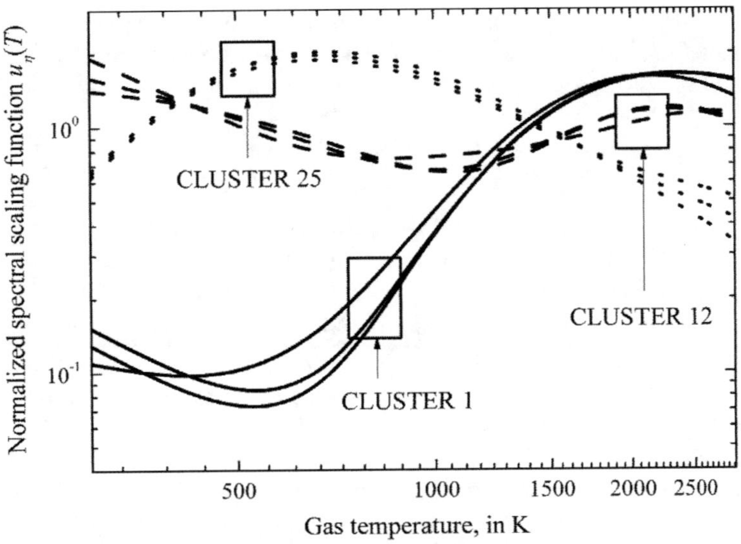

Figure 4. Examples of spectral scaling function for H_2O over the [3287.5 ; 3312.5] cm^{-1} spectral interval. Clusters were obtained with the functional form FF2.

Following Eqs. (32,38), on which the MSF is based, it is clear that the present approach shares many similarities with other existing models. In fact, any number of scaling functions can be used in Eq. (32). Let us thus consider a LBL spectrum calculated at high resolution and write N the number of points for which the absorption coefficient was evaluated inside $\Delta\eta$. Several cases can be considered:

- If $P = N$, then MSF model and LBL calculation are rigorously the same.
- If $P = 1$, then the combination of Eqs. (32,38) provides the so-called scaled-*k* model [14].
- If scaling functions are built in such a way that they associate the same value of $\hat{u}_p(\underline{\phi})$ to all the wavenumbers that correspond to a given value k_p of the absorption coefficient in the reference state, viz.: if $\kappa_\eta(\underline{\phi}^{ref}) = k_p \Rightarrow \hat{u}_p(\underline{\phi}) = \kappa_\eta(\underline{\phi}) / \kappa_\eta(\underline{\phi}^{ref}) = u_p(k_p, \underline{\phi})$, then we define a relationship between the gas spectra in the two states that can be written in the following form: $\kappa_\eta(\underline{\phi}) = u_p(k_p, \underline{\phi}) \cdot \kappa_\eta(\underline{\phi}^{ref}) = h\left[\kappa_\eta(\underline{\phi}^{ref})\right]$ where h is a function that correlates explicitly the values of the absorption coefficients in the two states: this provides exactly the correlated-*k* model [14]. However, no restriction about the behavior, monotonous or not, of the correlating function is then introduced.
- Theoretical links between MS and Fictitious gas approaches are discussed in Appendix.
- If sub-spectral intervals are built by grouping together wavenumbers associated to similar scaling functions, then: 1/ if scaling functions are defined explicitly by a given prescribed functional form, MSF provides the so-called multi-group approach described in Refs. [20,21] when the same basis of functions as proposed in these references are used, 2/ if those scaling functions are defined implicitly, viz. without explicit functional form but in terms of their overall behavior (ie. increase or decrease over the whole set of thermophysical states, or increase and then decrease, etc), the MSF provides the multi-group approach described in Refs. [22,23].
- If $P = 1$ and if we assume that the *k*-distribution function is Inverse Gaussian, we finally end up with the SNB Malkmus' model together with the Curtis-Godson approximation (this can be readily shown by considering the solutions of Eq. (38) for the scaling function at the optically thin and thick limits, that provide in this case analytical solutions, and then by reporting the results inside Eq. (32) written over a non-uniform path).

In other words, the set of Eqs. (32,38) provides the most general approximate model for gas radiation in non-uniform media. All existing models can thus be obtained by different ways to approximate the exact formula set by Eqs. (20,22).

5. Applications

5.1. Preliminary comments

All results provided in this section correspond to mixtures of H_2O and N_2. Reference LBL spectra were calculated using the HITEMP2010 [24] spectroscopic database for 27 temperature values between T_{min} = 300 K and T_{max} = 2,900 K and molar fractions x_{H_2O} = 0.01, 0.1, 0.2, 0.4, 0.6, 0.8 and 1.0. The total pressure is 1 atm. More details about those calculations can be found in Ref. [25]. The only differences in the present work are that: 1/ total partition sums were calculated using the polynomial formulas given in Ref. [26], 2/ the distances from line centers for the estimation of Lorentz profiles were taken from Ref. [27], 3/ the data used for the calculation of the Lorentz HWHM of the spectral lines where those given in the HITEMP2010 database.

Once LBL data are available, spectral groups (clusters) are built following the method described in section 4.2. (As said in section 4.3, clusters can be made using a reference pressure and composition. Here, values P^{ref} = 1 atm and $x^{ref}_{H_2O}$ = 0.1 were chosen). A detailed analysis of those sets of wavenumbers has shown that the clusters are mostly independent on the choice of the Functional Form FF1 or FF2. This was confirmed by several test cases (not given here) where the two models were found to provide results within ~0.1% on narrow band calculations. Accordingly, only the model based on FF2 will be considered here and will be referred to as MSC*k*.

Four approximate models are considered for comparisons with LBL reference data: C*k*-25, which is the usual narrow band C*k* model averaged over 25 cm^{-1} spectral intervals [1]; C*k*-1, the C*k* model with data generated at 1 cm^{-1}; MSC*k* with functional forms FF2 with p = 25 (distinct scaling functions); and *k*-Map (only for 0D test cases) viz. based on a direct application of Eqs. (20,21). To ensure fair comparison among the four models, results were averaged over 25 cm^{-1}. Notice that C*k*-1 and MSC*k* models share the same computational cost, but use different discretizations of the narrow bands.

The building of those model parameters was done by using the same method as described in Ref. [8]. This approach requires defining a reference temperature. Here we have chosen T^{ref} = 2,900 K (this value, together with P^{ref} = 1 atm and $x^{ref}_{H_2O}$ = 0.1 chosen previously, defines our reference state). 15 values of k^{ref} were used to discretize the interval $\left[k^{ref}_{\min}, k^{ref}_{\max}\right]$. Those minimum and maximum values of the absorption coefficient in the reference state were taken directly from the LBL dataset. Then, the corresponding values of absorption coefficient in all the other thermophysical states were obtained by solving Eq. (39).

In the next section, 0D test cases only involve two distinct thermophysical states. Notice that for these cases, *k*-Map uses 15^2 values of k. 1D calculations were performed using the semi-analytical method described in Ref. [28].

5.2. Examples of applications in 0D and 1D geometries

To illustrate the validity of the various models considered in this work, we first consider two simple 0D examples taken from Ref. [14]. In these cases, a slab of hot gas (at 1,000 K or 2,000 K) radiates through a cold slab at 300 K. Both layers have the same length (50 cm). The total pressure is 1 atm and the molar fractions of H_2O (x_{H_2O} = 0.2) are the same in both slabs. Results, in terms of emissivities as defined by Modest in Ref. [14], are depicted in figure 5 (T_h = 1,000 K) and figure 6 (T_h = 2,000 K). The C*k*-25 model provides in both cases the worst results, with errors that reach 55% when compared to the reference LBL data. Those results are consistent with those reported in Ref. [14]. The C*k*-1 model is more accurate, but errors up to 20% are observed. Significant improvements can be obtained with the MSC*k* and *k*-Map models for which errors do not exceed 4.2% for MSC*k* and 2.2% for *k*-Map in the worst cases. The *k*-Map approach is slightly more accurate than MSC*k* with a band averaged (over the 1000-2500 cm^{-1} spectral interval) error that is lower than 0.11%, compared to an error of up to 1.2% for MSC*k*.

The third test case was taken from Ref. [3]. It corresponds to a 1D triangular temperature profile: the temperature is constant and equal to the wall temperature T_w=500 K except at the centre of the domain where it increases up to 2,500 K inside a layer of width 10 cm. The gas is homogeneous (x_{H_2O} = 0.1) and the total pressure is 1 atm. Results for this case, which are provided in terms of the total net flux as a function of the distance between the walls, are depicted in figure 7. Again, the MSC*k* model clearly outperforms more conventional C*k* approaches. Furthermore, if we compare the order of magnitude of the errors on wall fluxes obtained here with those given in Ref. [3] (see figure 7 in this reference), it appears that the MSC*k* and Fictitious Gases (FG) approaches provide in this case very similar accuracies. This is not really surprising because, as shown for instance in Appendix, FG and MS techniques share many similarities. Nevertheless, one of the advantages of MSC*k* over CKFG is that CKFG approach is usually applied in terms of transmissivity (when formulated in terms of absorption coefficients, its computational cost is higher than a LBL calculation). Therefore, it cannot be used when the gas is surrounded by reflecting walls [3]. The MSC*k* method, on the other hand, does not suffer from the same possible source of error because, over clusters, the gas radiation model reduces to a usual correlated-*k* model, known to be applicable with non-black boundaries. Furthermore, for the same reason, MSC*k* can be applied in uniform media without any loss of accuracy, which may not be the case with the CKFG model (see for instance [3]).

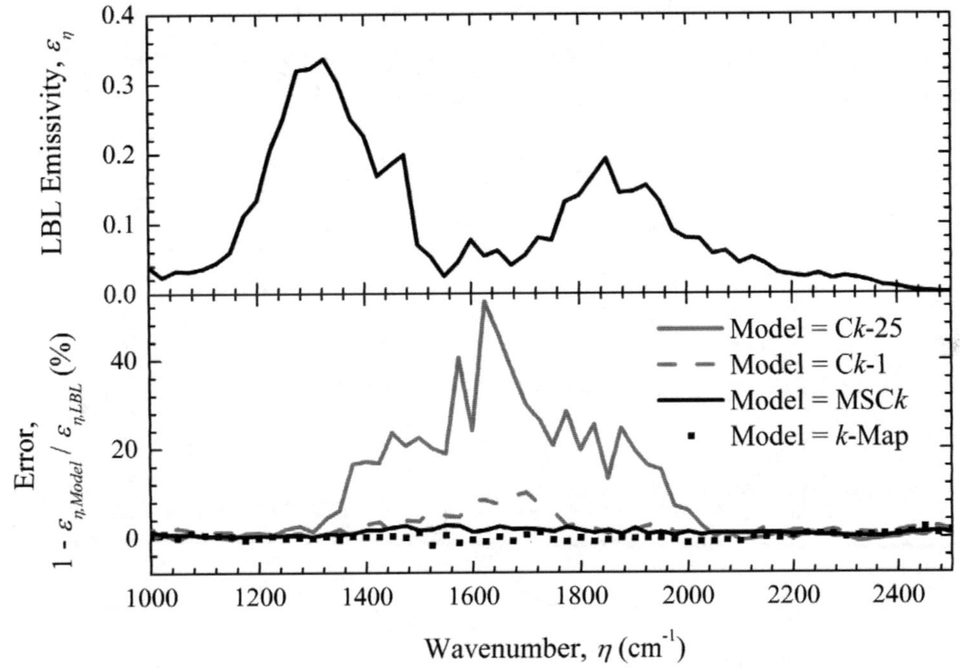

Figure 5. Narrow band emissivities for the 6.3 µm band of H_2O calculated by the LBL (top) method and errors due to the use of C*k*-25, C*k*-1, MSC*k* and *k*-Map approximate models. T_h = 1,000 K.

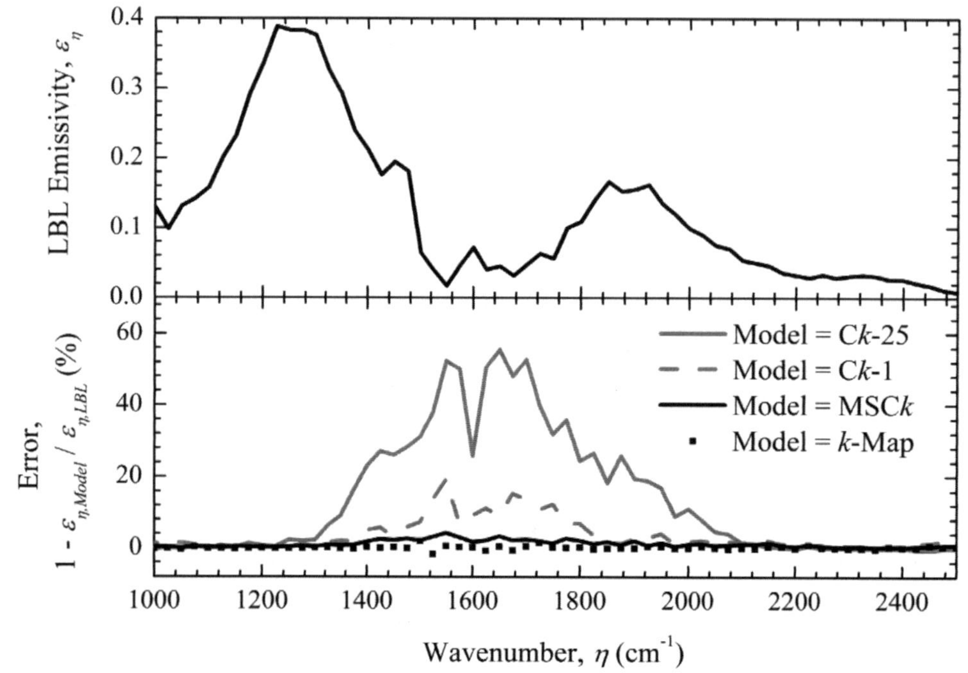

Figure 6. Narrow band emissivities for the 6.3 µm band of H_2O calculated by the LBL (top) method and errors due to the use of C*k*-25, C*k*-1, MSC*k* and *k*-Map approximate models. T_h = 2,000 K.

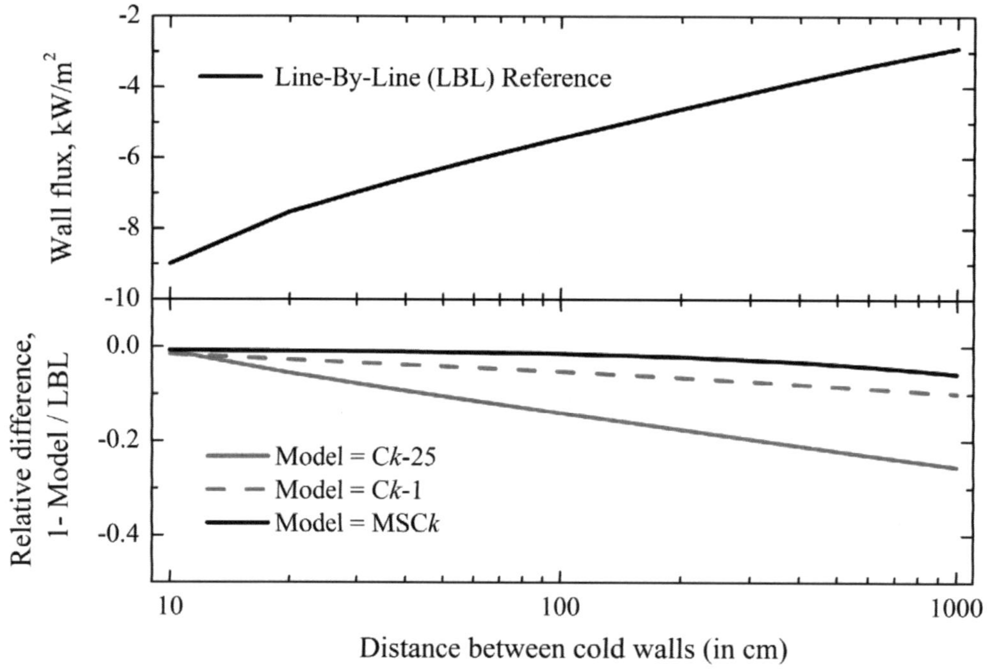

Figure 7. Wall net flux for a triangular temperature profile (from Ref. [3]) – the gas (H_2O-N_2 mixture, x_{H_2O} = 0.1, P = 1 atm) is surrounded by cold black walls at 500 K. Top: wall net flux calculated LBL, bottom: relative difference between model and LBL calculations. The hot layer thickness is 10 cm.

6. Conclusion

A general formula for the application of *k*-distribution approaches in non-uniform gaseous media was proposed, together with a simpler and computationally more efficient technique (based on the so-called Multi-Spectral Framework (MSF)) to allow its application in radiative heat transfer. Both models (the exact approach called *k*-Map, and its approximation, MSC*k*) were assessed against LBL reference data for H_2O-N_2 mixtures in 0D geometries. For 1D geometries, the exact approach cannot be recommended due to its computational cost, and in this case the use of the MSC*k* model is appropriate. Those two approaches were found to outperform usual *k*-distribution techniques in several test cases taken from the literature. Further developments, planned as future work, are still required to optimize the MSC*k* model (in terms of the number of values of *k* required for the calculation of *k*-integrals, and in terms of the optimal number of clusters) for application in radiative heat transfer. The application of the proposed method over the full spectrum will also be studied.

Acknowledgments
This work was supported by the French National Research Agency under the grant ANR-12-BS09-0018 SMART-LECT.

Appendix. Functional *k*-distribution formulation

We consider a narrow band $\Delta\eta$. Over this interval, the spectral absorption coefficient is the sum of the contributions of N_l spectral lines (whose centers may be inside or outside $\Delta\eta$). Each line corresponds to a radiative transition between two energy states of the molecules. Let us write E_l the lowest value of energy for the *l*-th spectral line. It is possible to assemble the spectral lines into N groups (equivalent to fictitious gases) with the same value of parameter E_l. If all spectral lines are associated to distinct values of E_l, then $N = N_l$. More generally, $N \le N_l$.

Let us assume that for any wavenumber $\eta \in \Delta\eta$, the spectral absorption coefficient of the gas in a thermophysical state given by the state vector $\underline{\phi}$ (temperature *T*, composition *x* and total pressure *P*) can be written as the sum of N scaled spectra, viz.

$$\kappa_\eta\left(\underline{\phi}\right)=\kappa_{\eta,1}\left(\underline{\phi}^{ref}\right)\cdot e_1'\left(\underline{\phi}\right)+\kappa_{\eta,2}\left(\underline{\phi}^{ref}\right)\cdot e_2'\left(\underline{\phi}\right)+\ldots+\kappa_{\eta,N}\left(\underline{\phi}^{ref}\right)\cdot e_N'\left(\underline{\phi}\right) \tag{A.1}$$

where $\underline{\phi}^{ref}$ represents a reference state and $e_i'\left(\underline{\phi}\right)$, $i=1,..,N$ the scaling functions. The corresponding temperature, T^{ref}, is chosen here equal to $T_{\min}$.

This statement is equivalent to assume that the profiles associated to the spectral lines that appear in the spectrum (that are here assumed to be Lorentzian) are scaled. This approximation was used in Ref. [6] to introduce the Fictitious gases approach. It was justified as follows: at low resolution, line-wings have a more important impact that line centers. Accordingly one can assume, as a first approximation, that line profiles are scaled viz.

$$\begin{aligned}
f_L\left(\eta-\eta_l,\underline{\phi}\right)&=\frac{\gamma_L\left(\underline{\phi}\right)}{\pi}\frac{1}{\left[\gamma_L\left(\underline{\phi}\right)\right]^2+\left(\eta-\eta_l\right)^2}=\frac{\gamma_L\left(\underline{\phi}\right)}{\gamma_L\left(\underline{\phi}^{ref}\right)}\frac{\gamma_L\left(\underline{\phi}^{ref}\right)}{\pi}\frac{1}{\left[\gamma_L\left(\underline{\phi}\right)\right]^2+\left(\eta-\eta_l\right)^2}\\
&\approx\frac{\gamma_L\left(\underline{\phi}\right)}{\gamma_L\left(\underline{\phi}^{ref}\right)}\frac{\gamma_L\left(\underline{\phi}^{ref}\right)}{\pi}\frac{1}{\left[\gamma_L\left(\underline{\phi}^{ref}\right)\right]^2+\left(\eta-\eta_l\right)^2}\\
&=\frac{\gamma_L\left(\underline{\phi}\right)}{\gamma_L\left(\underline{\phi}^{ref}\right)}f_L\left(\eta-\eta_l,\underline{\phi}^{ref}\right)
\end{aligned} \tag{A.2}$$

where η_l is the center of the l-th line, γ_L the half-width at half-maximum of the Lorentz profile assumed to be the same for all spectral lines, f_L the Lorentz profile and $\underline{\phi}$ the state vector as defined previously. γ_L is usually (see for instance Ref. [29]) estimated as the product of a function of temperature, $\gamma_{L,T}(T)$, and a function of composition, $\gamma_{L,c}(x,P)$:

$$\gamma_L(\underline{\phi}) = \gamma_{L,T}(T)\ \gamma_{L,c}(x,P) \tag{A.3}$$

Furthermore, we assume that: 1/ all spectral lines share the same partition function $Q(T)$, 2/ all terms of the form $1-exp(-hc\eta_l/k_BT)$ (that appear in the calculation of linestrengths) can be approximated as $1-exp(-hc\overline{\eta}/k_BT)$ where $\overline{\eta}$ represents the band center and k_B is the Boltzman's constant. Then we can write functions $e_i'(\underline{\phi})$, $i=1,..,N$ explicitly:

$$\begin{aligned} e_i'(\underline{\phi}) &= \underbrace{\frac{1-exp(-hc\overline{\eta}/k_BT)}{1-exp(-hc\overline{\eta}/k_BT^{ref})}\frac{Q(T^{ref})}{Q(T)}\frac{T^{ref}}{T}}_{v(T)}\underbrace{\frac{\gamma_{L,T}(T)}{\gamma_{L,T}(T^{ref})}}_{G_T(T)}\underbrace{\frac{\gamma_{L,c}(x,P)}{\gamma_{L,c}(x^{ref},P^{ref})}}_{G_c(x,P)}\underbrace{exp\left[-\frac{E_i}{k_B}\left(\frac{1}{T}-\frac{1}{T^{ref}}\right)\right]}_{e_i(T)} \\ &= v(T)\cdot G_T(T)\cdot G_c(x,P)\cdot e_i(T) \end{aligned} \tag{A.4}$$

As $v(T)G_T(T)\neq 0$, it is possible to reformulate Eq. (A.1) into:

$$\begin{cases} \kappa_\eta(\underline{\phi}) = \dfrac{\kappa_\eta(x^{ref},P^{ref},T)}{v(T)G_T(T)} \times v(T)G_T(T) \times G_c(x,P) \\ \dfrac{\kappa_\eta(x^{ref},P^{ref},T)}{v(T)G_T(T)} = \kappa_{\eta,1}(\underline{\phi}^{ref})\cdot e_1(T) + \kappa_{\eta,2}(\underline{\phi}^{ref})\cdot e_2(T) + \ldots + \kappa_{\eta,N}(\underline{\phi}^{ref})\cdot e_N(T) \end{cases} \tag{A.5}$$

where:

$$e_i(T) = exp\left[-\frac{E_i}{k_B}\left(\frac{1}{T}-\frac{1}{T^{ref}}\right)\right] \tag{A.6}$$

In Eq. (A.5), $\kappa_\eta(x^{ref},P^{ref},T)$ represents the spectral absorption coefficient at temperature T, composition x^{ref} and total pressure P^{ref}.

Now, let us consider the set of temperature $T_j, j=1,..,N$ defined as:

$$\frac{1}{T_j} - \frac{1}{T^{ref}} = \frac{j-1}{N-1}\left(\frac{1}{T_{\max}} - \frac{1}{T_{\min}}\right) \tag{A.7}$$

We can then use (A.5) to set the linear system of equations:

$$\boldsymbol{k} = J_\Phi \boldsymbol{K} \Leftrightarrow \boldsymbol{K} = J_\Phi^{-1}\boldsymbol{k} \tag{A.8}$$

where $\boldsymbol{k}$ and $\boldsymbol{K}$ are the vectors of components $[\boldsymbol{k}]_i = \kappa_\eta(x^{ref},P^{ref},T_i)/[v(T_i)G_T(T_i)]$ and $[\boldsymbol{K}]_i = \kappa_{\eta,i}(\underline{\phi}^{ref})$ respectively. J_Φ is a N by N square invertible matrix (with the set of temperatures given by Eq. (A.7), this matrix is of the Vandermonde type which is here invertible because all E_i's are distinct). Mathematically, Eq. (A.8) represents a change of basis and J_Φ is the corresponding

Jacobian matrix whose coefficients are: $\left[J_\Phi\right]_{ij} = \dfrac{\partial\left[\kappa_\eta\left(x^{ref}, P^{ref}, T_i\right)/v(T_i)G_T(T_i)\right]}{\partial\kappa_{\eta,j}\left(\underline{\phi}^{ref}\right)} = e_j(T_i)$. In this case, the set $e_1(T),..,e_N(T)$ defines a basis of a space $H(e_1,..,e_N)$ of functions $f(T)$ over the interval $T \in [T_{\min}, T_{\max}]$:

$$H(e_1,..,e_N) = \left\{ \begin{array}{c} \text{functions } \mathrm{K}: [T_{\min}, T_{\max}] \to \mathbb{R} \text{ such that} \\ \mathrm{K}(T) = v(T)G_T(T)\sum_{j=1}^{N} a_j\, e_j(T),\ \forall (a_1,..,a_N) \in \mathbb{R}^N \end{array} \right\} \tag{A.9}$$

Now, using Eq. (A.5), the band averaged transmission function of a path of length L inside the gas in any thermophysical condition $\underline{\phi}$ can be written as the multiple integral:

$$\tau^{\Delta\eta}(L) = \underbrace{\int_0^{+\infty}\int_0^{+\infty}...\int_0^{+\infty}}_{N \text{ integrals}} \exp\left\{-xP\, v(T)G_T(T)\, G_c(x,P)\left[K_1 \cdot e_1(T) + ... + K_N \cdot e_N(T)\right]L\right\} d\tilde{g}_N(K_1,..,K_N) \tag{A.10}$$

where:

$$\tilde{g}_N(K_1,..,K_N) = \frac{1}{\Delta\eta}\int_{\Delta\eta} H\left[K_1 - \kappa_{\eta,1}\left(\underline{\phi}^{ref}\right)\right] ... H\left[K_N - \kappa_{\eta,N}\left(\underline{\phi}^{ref}\right)\right] d\eta \tag{A.11}$$

Let us write β_{ij} the elements of the inverse of matrix J_Φ: $\left[J_\Phi^{-1}\right]_{ij} = \beta_{ij}$.

We have, following (A.8):

$$\begin{aligned} K_1 \cdot e_1(T) + ... + K_N \cdot e_N(T) &= \left(\sum_{i=1}^{N}\beta_{1i}k_i\right)\cdot e_1(T) + ... + \left(\sum_{i=1}^{N}\beta_{Ni}k_i\right)\cdot e_N(T) \\ &= k_1 \cdot \underbrace{\sum_{j=1}^{N}\beta_{j1}e_j(T)}_{\varepsilon_1(T)} + ... + k_N \cdot \underbrace{\sum_{j=1}^{N}\beta_{jN}e_j(T)}_{\varepsilon_N(T)} \end{aligned} \tag{A.12}$$

and, formally, from Ref. [13]:

$$\begin{aligned} \frac{\partial^N \tilde{g}_N(K_1,..,K_N)}{\partial K_1..\partial K_N} &= \frac{1}{\Delta\eta}\int_{\Delta\eta}\delta\left[K_1 - \kappa_{\eta,1}\left(T^{ref}\right)\right] ... \delta\left[K_N - \kappa_{\eta,N}\left(T^{ref}\right)\right] d\eta \\ &= \left|det\, J_\Phi\right| \frac{1}{\Delta\eta}\int_{\Delta\eta}\delta\left[k_1 - \frac{\kappa_\eta\left(x^{ref}, P^{ref}, T_1\right)}{v(T_1)G_T(T_1)}\right] ... \delta\left[k_N - \frac{\kappa_\eta\left(x^{ref}, P^{ref}, T_N\right)}{v(T_N)G_T(T_N)}\right] d\eta \\ &= \left|det\, J_\Phi\right| \frac{\partial g_N^v(k_1,..,k_N)}{\partial k_1...\partial k_N} \end{aligned} \tag{A.13-a}$$

where:

$$g_N^v(k_1,..,k_N) = \frac{1}{\Delta\eta}\int_{\Delta\eta} H\left[k_1 - \frac{\kappa_\eta\left(x^{ref}, P^{ref}, T_1\right)}{v(T_1)G_T(T_1)}\right] ... H\left[k_N - \frac{\kappa_\eta\left(x^{ref}, P^{ref}, T_N\right)}{v(T_N)G_T(T_N)}\right] d\eta \tag{A.13-b}$$

This yields:

$$\tau^{\Delta\eta}(L)=\underbrace{\int_0^{+\infty}\int_0^{+\infty}\cdots\int_0^{+\infty}}_{N\text{ integrals}} \exp\left\{-xPL\, v(T)G_T(T)\, G_c(x,P)\left[k_1\cdot\mathcal{E}_1(T)+\ldots+k_N\cdot\mathcal{E}_N(T)\right]\right\} dg_N^v(k_1,\ldots,k_N) \tag{A.14}$$

Accordingly, formulas (A.10) and (A.14) are rigorously equivalent. They can be both written formally in functional integral forms (so as to simplify the notations):

$$\begin{cases} \tau^{\Delta\eta}(L)=\displaystyle\int_{\mathrm{K}(T)\in H(e_1,\ldots,e_N)} \exp\left[-xPG_c(x,P)\ \mathrm{K}(T)L\right] dW\left[\mathrm{K}(T)\right] \\ \tau^{\Delta\eta}(L)=\displaystyle\int_{\kappa(T)\in H(\mathcal{E}_1,\ldots,\mathcal{E}_N)} \exp\left[-xPG_c(x,P)\ \kappa(T)L\right] dw\left[\kappa(T)\right] \end{cases} \tag{A.15}$$

where $dW\left[\mathrm{K}(T)\right]=d\tilde{g}_N(K_1,\ldots,K_N)$, $dw\left[\kappa(T)\right]=dg_N^v(k_1,\ldots,k_N)$ and where we define the following space of functions:

$$H(\mathcal{E}_1,\ldots,\mathcal{E}_N)=\left\{\begin{matrix}\text{functions } \kappa:\left[T_{\min},T_{\max}\right]\to\mathbb{R} \text{ such that} \\ \kappa(T)=v(T)G_T(T)\cdot\sum_{j=1}^{N}a_j\mathcal{E}_j(T),\ \forall(a_1,\ldots,a_N)\in\mathbb{R}^N\end{matrix}\right\} \tag{A.16}$$

In Eq. (A.15), $dW\left[\mathrm{K}(T)\right]=d\tilde{g}_N(K_1,\ldots,K_N)$ defines a measure on the space of functions $H(e_1,\ldots,e_N)$. Similarly, $dw\left[\kappa(T)\right]=dg_N^v(k_1,\ldots,k_N)$ is a measure on $H(\mathcal{E}_1,\ldots,\mathcal{E}_N)$. Eq. (A.15) can be compared to the more usual formulas Eqs. (1,2). That becomes what can be called a functional *k*-distribution formulation.

Formulas (A.15) can be obviously extended to non-uniform media. Nevertheless, those formulations do not have (*a priori*) a real benefit over Eq. (20) for applications in radiative heat transfer. Indeed, most problems are typically formulated in terms of discretized temperature fields (viz. values of temperatures at discrete points sampled through a mesh). In those applications, Eq. (20) provides the exact solution. Nevertheless, this formulation is interesting from a fundamental point of view because:

- it noticeably exhibits the links that exist between *k*-distributions in non-uniform media and functional data concepts and methods (such as FDA), as used in the multispectral approach.
- it clearly shows that LBL and Fictitious gas methods (which are mainly founded on the first integral formulation in Eq. (A.15)) and mapping techniques (based on the second formula in the same set of equations) are rigorously equivalent.

Finally, it should be noticed that the problem usually encountered for the definition of a measure in functional or path integration (see for instance the discussion pages 4-5 in Ref. [30] - the same problem is found in Functional Data Analysis, [31]) is avoided here because the spaces of functions considered for the calculation of the integrals are of finite dimensions N (that represents the number of spectra that can be associated to distinct values of E_i that one can find in the spectroscopic database - this number is always finite). Over those spaces, a measure can be defined in terms of the components of any function in the basis $e_1(T),\ldots,e_N(T)$ or $\mathcal{E}_1(T),\ldots,\mathcal{E}_N(T)$. This may not be the case in a general frame [32].

References

[1] Modest MF 2003 *Radiative heat transfer* (New York: Academic Press)
[2] Goody RM, Yung YL 1989 *Atmospheric radiation - theoretical basis* (New York: Oxford University Press)

[3] Pierrot L, Rivière Ph, Soufiani A, Taine J 1999 *J. Quant. Spectr. Rad. Transfer* **62** 523-548
[4] Modest MF, Zhang H 2003 *J. Heat Transfer* **124** 30-38
[5] Denison MK, Webb BW 1993 *J. Heat Transfer* **115** 1004-1012
[6] Levi di Leon R, Taine J 1986 *Rev. Phys. Appl.* **21** 825-831
[7] West R, Crisp D, Chen L 1990 *J. Quant. Spectr. Rad. Transfer* **43** 191-199
[8] André F, Hou L, Roger M, Vaillon R 2014 *J. Quant. Spectr. Rad. Transfer* **147** 178-195
[9] Ambartzumian V 1934 *Publ Obs Astron Univ Leningrad* **6** 7-18
[10] Lebedinski AI 1939 *Proc Leningrad Univ Ser Math* **3(31)** 152-75
[11] Kondratiev KY 1969 *Radiation in the atmosphere* (New York: Academic Press)
[12] Arking A, Grossman K 1972 *J Atmos Sci* **29** 937-949
[13] Kanwal R 1998 *Generalized functions: theory and practice* (Boston: Birkhäuser)
[14] Modest MF 2003 *J. Quant. Spectr. Rad. Transfer* **76** 69-83
[15] Abramowitz M, Stegun IA 1965 *Handbook of Mathematical Functions* (Dover)
[16] Everitt BS, Landau S, Leese M, Stahl D 2011 *Cluster analysis* (John Wiley and Sons)
[17] Ramsay JO, Silverman BW 2005 *Functional data analysis* (Springer series in statistics)
[18] Jacques J, Preda C 2013 *Functional data clustering: a survey* Research Report N°8198
[19] Ferreira L, Hitchcock DB 2009 *Communications in Statistics* **38** 1925-1949
[20] Zhang H, Modest MF 2002 *J. Quant. Spectr. Rad. Transfer* **73** 349-360
[21] Zhang H, Modest MF 2003 *J. Heat Transfer* **125** 454-461
[22] Pal G, Modest MF, Wang L *J. Heat Transfer* **130** 082701-082708
[23] Pal G, Modest MF *J. Heat Transfer* **132** 023307-1-023307-9
[24] Rothman LS, et al 2010 *J. Quant. Spectr. Rad. Transfer* **111** 2139-2150
[25] Andre F, Vaillon R 2010 *J. Quant. Spectr. Rad. Transfer* **111** 1900-1911
[26] Vidler M, Tennyson J 2000 *J Chem Phys* **113** 9766-9771
[27] Riviere P, Soufiani A 2012 *Int. J. Heat Mass Transfer* **55** 3349-3358
[28] Solovjov VP, Webb BW 2008 *J. Quant. Spectr. Rad. Transfer* **109** 245-257
[29] Rothman LS, et al 1998 *J. Quant. Spectr. Rad. Transfer* **60** 665-710
[30] Chaichian M, Demichev A 2001 *Path integrals in Physics Volume I Stochastic processes and quantum mechanics* (Bristol and Philadelphia: IOP)
[31] Delaigle A, Hall P 2010 *Annals Stat* **38** 1171-1193
[32] Menskii MB 1992 *Theoretical and Math Phys* **93** 1262-1267

Specular reflective boundary conditions for Discrete Ordinate Methods in Periodic or Symmetric Geometries

Jian Cai[*], **Michael F. Modest**[†]

[*] Assistant Professor, University of Wyoming, Laramie, WY 82071, USA
[†] Shaffer and George Professor of Engineering, University of California, Merced, CA 95343, USA

E-mail: jcai@uwyo.edu, mmodest@ucmerced.edu

Abstract. In simulations of periodic or symmetric geometries, computational domains are reduced by imaginary boundaries that present the symmetry conditions. In Photon Monte Carlo methods, this is achieved by imposing specular reflective boundary conditions for the radiative intensity. In this work, a similar specular reflective boundary condition is developed for Discrete Ordinate Methods. The effectiveness of the new boundary condition is demonstrated by multiple numerical examples including plane symmetry and axisymmetry.

1. Introduction

Periodic and symmetric geometries are frequently encountered in engineering applications. Examples of such geometries include periodic structures, plane symmetry and axisymmetry. In the simulations of these problems, flow variables are solved for only a section of the domain. The computational domain is separated from the rest by imaginary boundaries, upon which symmetric constraints as opposed to physical conditions are applied.

While the symmetry constraints may be easily expressed into mathematical formulas for scalar and vector fields as frequently performed in CFD simulations, they present challenges for radiative intensities. This is because the radiative intensities are functions of both spatial and directional coordinates [1]. Plane and rotational symmetric conditions are complicated by the additional directional variations. As an example, scalar flow variables in an axisymmetric problem are only functions of radial and axial locations that fit into two-dimensional cylindrical coordinates. In flow solvers designed for only three-dimensional coordinates, a wedge geometry that has only one layer of cells in the azimuthal direction may be employed. Symmetry conditions are imposed on the imaginary wedge planes. However, even in these axisymmetric problems, the radiative intensities are still three-dimensional, i.e., light propagation is not limited to the plane formed by radial and axial direction vectors. Such issues are also observed in Photon Monte Carlo (PMC) methods for radiative transfer. In PMC methods, the wedge planes may be treated as specular reflective boundary conditions, i.e., a ray hitting the wedge plane is reflected back into the computational domain like hitting a perfect mirror.

When applying Discrete Ordinate Methods [1, 2] to axisymmetric simulations, three methods are typically employed. In the first method, the Radiative Transfer Equations (RTE) are solved for ordinate intensities in a full three-dimensional axisymmetric domain, as in Ref. [3, 4]. This method cannot take advantage of symmetry to reduce computational cost. In the second method, the RTEs are formulated in cylindrical coordinates directly as in Ref. [5–7]. However, this approach requires a different implementation of the equations. In the third method, a reduced mesh based on symmetry is used to discretize the RTE formulated for Cartesian coordinates [8, 9]. Axisymmetry is enforced by special treatment of the wedge boundaries. This method simplifies the implementations of RTEs, because the same solver and discretization method are used for both reduced geometry or the full three-dimensional geometry. However, the resulting new wedge boundaries require different treatment from that employed for scalar or vector partial differential equations.

In this work, a specular reflective boundary condition is proposed for Discrete Ordinate Methods in recognition of the fact that the ordinates have similarities to rays in Photon Monte Carlo methods. The specular reflective boundary conditions are expressed as mathematical boundary conditions for the first-order partial differential equations that govern spatial variations of each ordinate depending on the spatial relationship between the surface normal and the ordinate directions. The proposed method makes a theoretical contribution to radiation solution methods and provides potential reduction in computational cost for symmetric geometries.

2. Theoretical background

2.1. Discrete Ordinate Methods

In this section, the Discrete Ordinate Methods for solving Radiative Transfer Equation are briefly reviewed. The presentation focuses only on necessary content for the new development as opposed to completeness. Discussions will be limited to nonscattering participating media, because the treatment of scattering is not essential to this work and does not impose new technical difficulties. Readers are referred to [1] for more comprehensive discussions and [2] for a review of recent developments.

The Radiative Transfer Equation (RTE) for a radiatively participating medium with emission and absorption is a first order differential equation:

$$\hat{\boldsymbol{s}} \cdot \nabla_{\boldsymbol{x}} I(\boldsymbol{x}, \hat{\boldsymbol{s}}) = \kappa I_{\rm b}(\boldsymbol{x}) - \kappa I(\boldsymbol{x}, \hat{\boldsymbol{s}}) \tag{1}$$

where $\boldsymbol{x}$ is the spatial coordinate, $\hat{\boldsymbol{s}}$ is a unit direction vector, κ is the absorption coefficient, I the radiative intensity and $I_{\rm b}$ the blackbody intensity (or Planck function). The subscript $\boldsymbol{x}$ of the gradient operator ∇ emphasizes that the gradient is respect to spatial coordinates only.

In Discrete Ordinate Methods the directional variation of the radiative intensity is expressed by intensities on a set of prescribed directions, known as the ordinates. For each of the n ordinates ($\hat{\boldsymbol{s}}_i, i = 1, \ldots, n$), the corresponding intensity ($I_i = I(x, \hat{\boldsymbol{s}}_i)$) is determined by solving the RTE for direction $\hat{\boldsymbol{s}}_i$, i.e.,

$$\hat{\boldsymbol{s}}_i \cdot \nabla I_i(x) = \kappa I_{\rm b}(x) - \kappa I_i(x) \tag{2}$$

The resulting RTE is a first order partial differential equation and depends on spatial coordinates only. Each ordinate $\hat{\boldsymbol{s}}_i$ has a directional quadrature weight w_i such that a directional integral is converted into a sum over quadratures. In particular,

$$\int_{\Omega} d\Omega = \sum_i w_i = 4\pi \tag{3}$$

$$\int_{\Omega} I d\Omega = \sum_i I_i w_i = G \tag{4}$$

where G is incident radiation.

2.2. Finite Volume Discretization

Equation (2) may be solved numerically by Finite Volume Methods, i.e., it is integrated within a cell before the Gauss theorem is used to convert differential operators into algebraic operations. For example, consider a cell C with a volume V_c enclosed by m_f faces (Fig. 1). The RTE, Eq. (2), is integrated within this cell according to

$$\int_{V_c} \hat{\boldsymbol{s}}_i \cdot \nabla I_i(x)\, dV = \int_{V_c} \kappa I_{\rm b}(x) dV - \int_{V_c} \kappa I_i(x) dV. \tag{5}$$

For the right-hand side, the cell-center values of radiative intensity (I_{ic}) and properties (κ_c) are defined such that

$$\int_{V_c} \kappa I_{\rm b}(x) dV - \int_{V_c} \kappa I_i(x) dV = V_c \kappa_c (I_{bc} - I_{ic}) \tag{6}$$

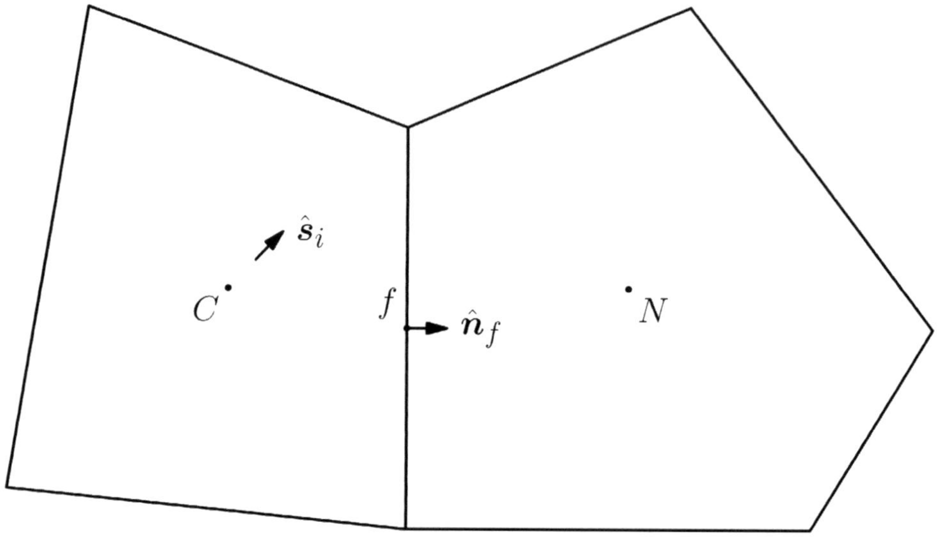

Figure 1. Geometry for convection schemes.

where I_{bc} is the Planck function evaluated at the cell-valued thermodynamic state.

The left hand side is related to I_i at the faces according to the Gauss theorem

$$\int_{V_c} \hat{\mathbf{s}}_i \cdot \nabla I_i(x)\, dV = \int_{S_c} I_i(x)\hat{\mathbf{s}}_i \cdot \hat{\boldsymbol{n}}_f dA = \sum_f \int_{A_f} I_i(x)\hat{\mathbf{s}}_i \cdot \hat{\boldsymbol{n}}_f dA = I_{if}\hat{\mathbf{s}}_i \cdot \hat{\boldsymbol{n}}_f A \tag{7}$$

where I_{if} is the intensity at face f, A_f the surface area of face f, and $\hat{\boldsymbol{n}}_f$ the surface normal of face f.

Further development is now required to express the face values I_{if} in terms of I_i. When the face f is an internal face, I_{if} is usually expressed as a function of I_i in the current cell (I_{iC}) and that in the neighbor cell (I_{iN}) that shares face f

$$I_{if} = F(I_{iC}, I_{iN}, \hat{\mathbf{s}}_i, \hat{\boldsymbol{n}}_f). \tag{8}$$

The simplest scheme is an upwind scheme known as the step scheme, i.e.,

$$I_{if} = F(I_{iC}, I_{iN}, \hat{\mathbf{s}}_i, \hat{\boldsymbol{n}}_f) = \begin{cases} I_{iC}, & \text{if } \hat{\mathbf{s}}_i \cdot \hat{\boldsymbol{n}}_f \geq 0 \\ I_{iN}, & \text{otherwise.} \end{cases} \tag{9}$$

However, when face f is on a computational boundary, a boundary condition is needed to determine the value of I_i on an ordinate $\hat{\mathbf{s}}_i$ outgoing from the face. For physical boundaries with prescribed radiative properties, the intensity from the boundary is calculated from temperature, emittance and incoming intensities. For imaginary computational boundaries, a specular reflective boundary condition is commonly used with Photon Monte Carlo methods.

2.3. Specular reflective boundary condition

Light is reflected on a specular reflective boundary as if on a mirror, as shown in Fig. 2. The top half is within the computational domain, and is separated from the imaginary domain by a specular reflective boundary (OO'). The reflected direction ($\boldsymbol{s}_r$) of an incoming direction ($\boldsymbol{s}_i$) after hitting a face with a normal direction $\hat{\boldsymbol{n}}_f$ is

$$\hat{\mathbf{s}}_r = 2\left(\hat{\boldsymbol{n}}_f \cdot \hat{\mathbf{s}}_i\right)\hat{\boldsymbol{n}}_f - \hat{\mathbf{s}}_i \tag{10}$$

Because of the reflective symmetry imposed by boundary, the intensities of the boundary cell (C) and its image (C') are related by

$$I_{i'C'} = I_{iC} \tag{11}$$

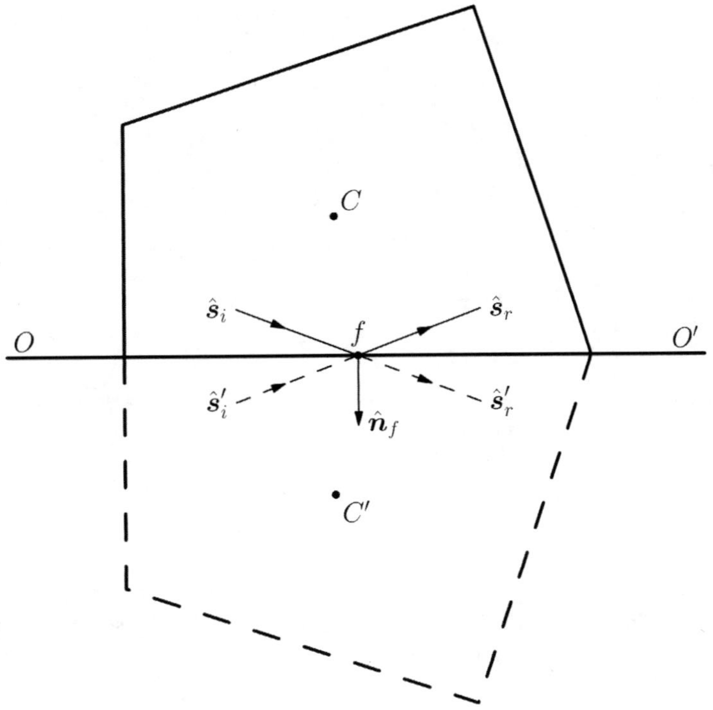

Figure 2. Reflective boundary condition. The boundary OO' separates computational domain (shown above in solid lines) from its image (shown below in dashed lines). The cell centers are C and its image C'. An incoming ordinate ($\hat{\mathbf{s}}_i$) is reflected back to computational domain as $\hat{\mathbf{s}}_r$. $\hat{\mathbf{s}}'_i$ and $\hat{\mathbf{s}}'_r$ are the images of the ordinates $\hat{\mathbf{s}}_i$ and $\hat{\mathbf{s}}_r$, correspondingly.

and the geometric constraints $\hat{\mathbf{s}}'_i = \hat{\mathbf{s}}_r$.

A successful reflective boundary condition should reproduce the same surface value as the internal surface when the full geometry including the image is employed. In other words, this requires that for any outgoing ordinate $\hat{\mathbf{s}}_r$, its intensity on the surface reads

$$I_{rf} = F(I_{rC}, I_{i'C'}, \hat{\mathbf{s}}_r, \hat{\boldsymbol{n}}_f) = F(I_{rC}, I_{iC}, \hat{\mathbf{s}}_r, \hat{\boldsymbol{n}}_f) \tag{12}$$

Notice that Eq. (11) is employed to replace the cell-valued intensity in imaginary cells ($I_{i'C'}$) with the intensity in boundary cells (I_{rC}).

Similarly, the equation for the boundary condition of the incoming ordinate $\hat{\mathbf{s}}_i$ can be derived,

$$I_{if} = F(I_{iC}, I_{r'C'}, \hat{\mathbf{s}}_i, \hat{\boldsymbol{n}}_f) = F(I_{iC}, I_{rC}, \hat{\mathbf{s}}_i, \hat{\boldsymbol{n}}_f) \tag{13}$$

When the simplest step (upwind) scheme (Eq. 9) is used, the above boundary conditions, Eqs. (12) and (13) can be simplified to

$$I_{rf} = I_{iC} \tag{14}$$

$$I_{if} = I_{iC} \tag{15}$$

These two equations suggest that when the step scheme is used, the boundary condition for incoming ordinates is zero gradient, while the boundary condition for the outgoing ordinate is fixed value of the cell value from the incoming ordinate before reflection. These simplified conditions for step scheme were also mentioned by [9]. However, for higher order spatial schemes, the boundary conditions are more complicated and involve cell values of both incoming and outgoing ordinates, as shown in Eqs. (12) and (13), respectively.

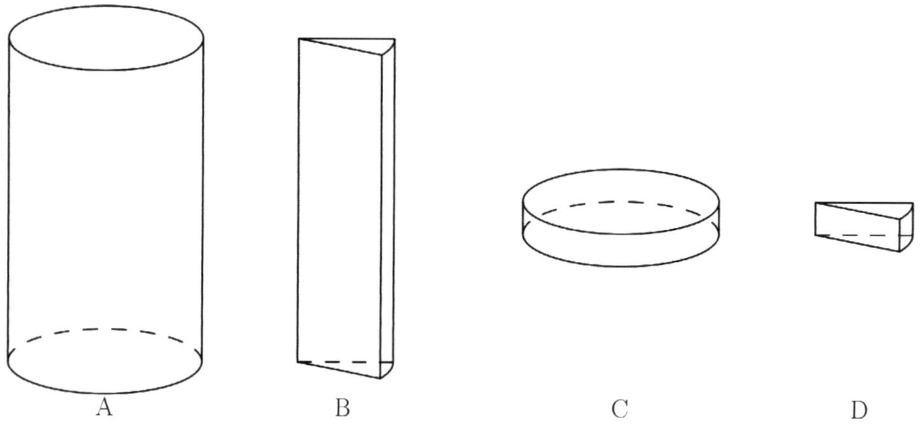

Figure 3. Four geometries.

3. Results and discussion

In this section sample calculations are carried out to demonstrate the effects of specular reflective boundary conditions on Discrete Ordinate Methods solutions. The boundary condition may be used for periodic or symmetry geometries. In the first example, the consistency is demonstrated.

3.1. An axisymmetric and periodic example

An axisymmetric radiation problem is considered here to demonstrate that the proposed boundary condition produces identical results as those from the full geometry. A homogeneous emitting–absorbing medium is confined in an infinitely long cylindrical enclosure bounded by black and cold walls. The problem is one-dimensional in the radial direction, and has two symmetry conditions. One is along the azimuthal direction, the other one is along the axial direction. The problem therefore may be solved with four different geometries as shown in Fig. 3. Geometry A is a sufficiently long cylinder bounded by black and cold walls at both ends. The radial cross section at the center achieves the target solution asymptotically as the length of the cylinder increases. Specularly reflective boundary conditions, therefore, are not needed in this geometry. Geometry B has a wedge shape with one layer of cells along the azimuthal direction, and takes advantage of the axisymmetry of the problem. Both of the additional front and back surfaces require the specular reflective boundary condition. Geometry C has a pie shape with only one layer of cells along the axial direction. It utilizes axial periodicity, such that both up and down surfaces are specularly reflective. Finally, Geometry D utilizes both axial periodicity and axisymmetry, and has cells only along the radial direction. Specularly reflective boundaries are applied to front, back, top and bottom surfaces. Because the solution (e.g., incident radiation or radiative heat source) of this problem has only radial variations, a successful boundary condition for the reduced geometries should produce identical results in all four geometries.

In all four geometries, a cylindrical grid with uniform spacing along axial, radial and azimuthal directions is used to discretize the RTE. The wedge geometries (Geometry B and D) have a 10° angle. Geometry A and C have 36 azimuthal cells with 10° for each wedge. Geometry A and B have 100 uniform axial cells, and Geometry C and D have identical height corresponding to the cell height in Geometry A and B. A total of 72 ordinates (2 polar and 36 azimuthal directions) are employed. The choice of large number of azimuthal direction ensures that the reflected ordinate is within the set of ordinates. The radial direction is uniformly discretized into 1000 cells. The radius of the cylinder is 1 m, and the absorption coefficient is 1 cm^{-1}; emission is homogeneous.

The solutions for normalized incident radiation ($G/4\pi I_b$) are compared in Fig. 4 with the exact results. Results from all four geometries overlap with each other, demonstrating the effectiveness of the proposed specular reflective boundary condition in dealing with both axisymmetric and periodic symmetries. The computational costs are 16720, 158, 15, and 2 seconds for Geometry A to D, respectively. A significant reduction of computational cost is achieved.

Journal of Physics: Conference Series **676** (2016) 012002 doi:10.1088/1742-6596/676/1/012002

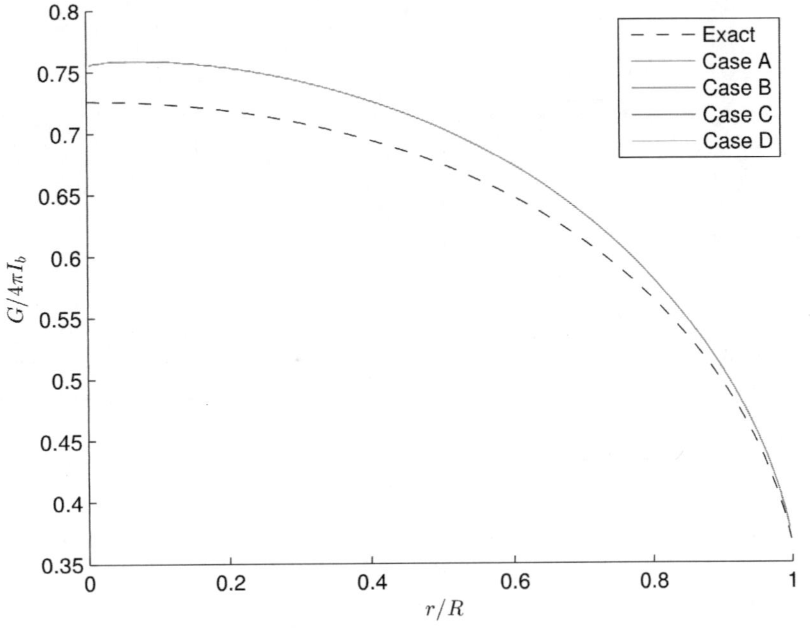

Figure 4. Comparison of the normalized incident radiation predicted from four geometries.

3.2. *Axisymmetric jet flame*

The specular reflective boundary condition is further applied to an axisymmetric jet diffusion flame solved on a wedge mesh similar to Geometry B in Fig. 3. The flame was derived from the methane–air partially premixed Sandia Flame D [10] by artificially quadrupling the jet diameter [11, 12] to make radiation in this small laboratory-scale flame appreciable. The wedge mesh has a 10° opening angle. Time-averaged profiles are used for the radiation calculations. The same flame was previously employed to investigate the accuracy and efficiency of k-distribution methods and P_N RTE solvers [13]. In this section the results from Discrete Ordinate Methods are reported. The spectral model is the full-spectrum k-distribution method assembled from a narrowband k-distribution database [14]. The polar direction is aligned with the axial direction. 36 azimuthal ordinates are employed so that reflected ordinates are also contained within the set of ordinates; the numbers of polar directions vary from 2 to 32.

The radiative heat flux divergence ($\nabla \cdot \boldsymbol{q}$), i.e., the negative of the radiative heat source in the energy equations predicted by different DOM implementations are compared with PMC and P_1 at 1 m and 1.43 m above the nozzle in Figs. 5 and 6, respectively. Radiative emission is determined by the Optically Thin approximation (OT). The true radiative heat sources are determined by PMC coupled with a line-by-line spectral model. Identical trends are found at both downstream locations. The P_1 model recovers most of the non-gray self-absorption when coupled with the full-spectrum k-distributions. DOM with 2 polar ordinates overpredicts absorption, increasing the number of polar ordinates improves the accuracy only slightly, and the amount of improvement is small compared to the overall amount of self-absorption. Therefore, DOM results from different ordinate configurations appear very close.

3.3. *Conclusions*

In this work, a specular reflective boundary condition similar to that employed in Photon Monte Carlo method is proposed to reduce computational geometry by exploring periodicity or symmetry of the problem. The expression of the boundary condition depends on the spatial discretization scheme for the divergence term. For the simplest step scheme, the specular reflective boundary condition for ordinate intensity reduces to zero gradient (or fixed value) for incoming (or outgoing) directions relative to the boundary. Therefore, the intensity of outgoing ordinate at the boundary after reflection depends on the intensity of the incoming ordinate, but not vice versa. However, for higher order schemes the incoming

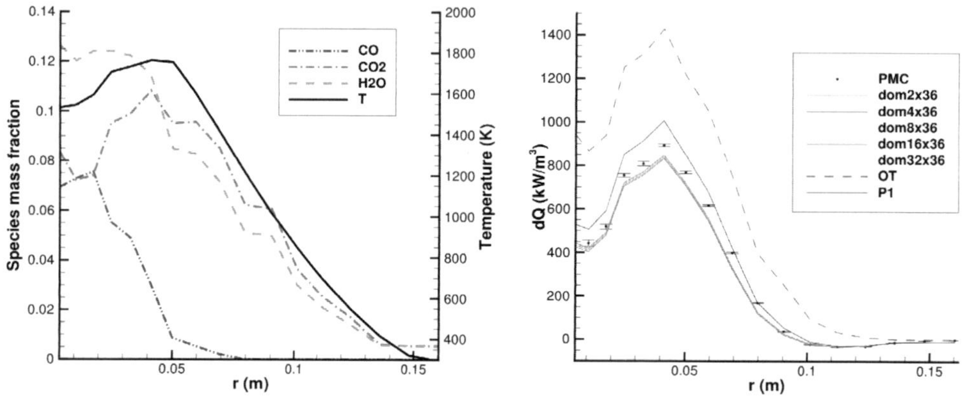

Figure 5. Radial scalar profiles (left) and prediction of radiative heat flux divergence (right) at 1 m above nozzle.

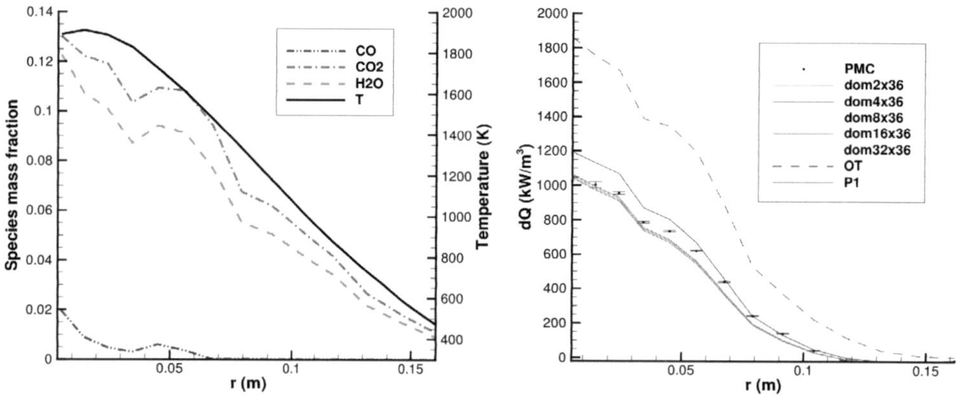

Figure 6. Radial scalar profiles (left) and prediction of radiative heat flux divergence (right) at 1.43 m above nozzle.

and outgoing ordinates are fully coupled. The consistency is demonstrated by a one-dimensional radial problem in a cylindrical coordinate that posses both axisymmetry and plane symmetry. The numerical example of a nongray axisymmetric jet flame solved on a wedge geometry with both radial and axial inhomogeneities is also provided.

[1] Modest M F 2013 *Radiative Heat Transfer* 3rd ed (New York: Academic Press)
[2] Coelho P J 2014 *Journal of Quantitative Spectroscopy and Radiative Transfer* **145** 121 – 146
[3] Chui E H, Raithby G D and Hughes P M J 1992 *Journal of Thermophysics and Heat Transfer* **6** 605–611
[4] Moder J P, Chai J C, Parthasarathy G, Lee H S and Patankar S V 1996 *Numerical Heat Transfer – Part B: Fundamentals* **30** 437–452
[5] Baek S W and Kim M Y 1997 *Numerical Heat Transfer – Part B: Fundamentals* **31** 313–326
[6] Fiveland W A 1982 ASME Paper 82-HT-20
[7] Murthy J Y and Mathur S R 1998 *Numerical Heat Transfer – Part B: Fundamentals* **33** 397–416
[8] Mathur S R and Murthy J Y 1999 *ASME Journal of Heat Transfer* **121** 357–364
[9] Kumar P and Eswaran V 2013 *ASME Journal of Heat Transfer* **12** 124501
[10] Barlow R S and Frank J H 1998 *Proceedings of the Combustion Institute* **27** 1087–1095
[11] Li G and Modest M F 2003 *ASME Journal of Heat Transfer* **125** 831–838
[12] Wang A, Modest M F, Haworth D C and Wang L 2008 *Journal of Quantitative Spectroscopy and Radiative Transfer* **109** 269–279
[13] Cai J, Marquez R and Modest M F 2014 *ASME Journal of Heat Transfer* **136** 112702
[14] Cai J and Modest M F 2014 *Journal of Quantitative Spectroscopy and Radiative Transfer* **141** 65–72

Methodology for optical characterization of multi-scale morphologically complex heterogeneous media – Application to snow with soot impurities

X Dai[1], S Haussener[1]
[1] Ecole Polytechnique Fédérale de Lausanne, Institute of Mechanical Engineering, Lausanne, Switzerland

E-mail: sophia.haussener@epfl.ch

Abstract. The radiative characterization of heterogeneous media with complex morphologies on multiple scales is of interest in a variety of areas such as solar energy conversion technologies or environmental sciences. An in-depth understanding and decoupling of the multi-scale morphological effects, bulk material properties, and operating conditions on the macroscopic behaviour provides pathways for morphology tailoring on multiple scales for improved application performance. We introduce a multi-scale methodology for the characterization of the spectral radiative transport in heterogeneous media with complex morphologies on two distinct scales characterized by size parameters (π·diameter/wavelength) significantly above and below one. The methodology incorporates the exact morphology at the various scales and utilizes volume-averaging approaches with the corresponding effective properties to couple the scales. At the large scale the volume-averaged coupled radiative transfer equations are solved utilizing *i*) effective radiative transport properties obtained by direct Monte Carlo simulations at the middle scale (mm range), and *ii*) averaged bulk material properties obtained at the small scale (submicron scale) by discrete dipole approximation calculations. The method is exemplary applied to snow containing agglomerated soot impurities. A quantification and decoupled understanding of the morphological effect on the radiative transport is achieved and a significant influence of the dual-scale morphology on the macroscopic optical behaviour is observed.

1. Introduction

Radiative transfer in heterogeneous media is observed in a variety of significant technical applications such as in solar energy conversion processes functioning as high temperature solar absorbers or photovoltaics devices, or in naturally occurring systems such as snow pack or porous soil. The maximization of the absorption in solar energy conversion devices or the accurate characterization of the reflection behavior of earth surfaces for climate modelling are the objectives and have particular relevance in today's society. The radiative characterization of such media is challenging as the exact incorporation of the real morphology is needed for an accurate radiative characterization. Tomography-based approaches have recently been successfully used to provide the detailed morphological information for radiative characterization down to a scale of micrometers [6]. Volume-averaging approaches provide a theoretical basis for the radiative characterization in such complex heterogeneous media through the introduction of effective transport properties, which account for the detailed morphology [9]. An additional complexity is added to radiative transport problems in heterogeneous media when the morphologies exhibit morphological details at various scales. We provide a methodology for the investigation of such media exhibiting morphological details at two distinct scales described by their size parameters ($x=\pi\cdot$diameter/wavelength) significantly above and below one.

2. Methodology

The methodology accounts for three scales described as small ($x<1$), middle ($x>1$), and large scale ($x>>1$), and consists of the following steps: *i*) obtaining the detailed 3D morphology of the sub-micron scaled structures ($x<1$) and its incorporation into the solution of the Maxwell's equations for the determination of the extinction- and scattering-efficiencies and scattering phase function, *ii*) obtaining the detailed 3D morphology of the micron sizes structures ($x>1$) and its incorporation into the solution of the discrete-scale radiative transfer equations (RTEs) for the determination of the effective extinction- and scattering coefficients and effective scattering phase functions, and *iii*) incorporation of the two-scale radiative transport properties into the coupled multi-phase volume-averaged RTEs for the determination of the averaged intensity vector field and heat transfer fluxes.

On the large-scale (continuum-scale) the intensity vector field within a heterogeneous medium composed of two different phases or components, is given by two volume-averaged RTEs [9]:

$$\begin{aligned}\hat{\mathbf{s}}\cdot\nabla I_{i,\lambda}(\mathbf{x},\hat{\mathbf{s}}) = & -\left[\kappa_{\mathrm{d},i,\lambda}+\sigma_{\mathrm{s,d},i,\lambda}+\sigma_{\mathrm{s,int},ii,\lambda}+\sigma_{\mathrm{s,int},ij,\lambda}\right]I_{i,\lambda}(\mathbf{x},\hat{\mathbf{s}})+n_{i,\lambda}^2\kappa_{\mathrm{d},i,\lambda}I_{\mathrm{b},i,\lambda}(\mathbf{x}) \\ & +\frac{1}{4\pi}\int_{4\pi} I_{i,\lambda}(\mathbf{x},\hat{\mathbf{s}}_{\mathrm{in}})\left[\sigma_{\mathrm{s,d},i,\lambda}\Phi_{\mathrm{d},i,\lambda}(\hat{\mathbf{s}}_{\mathrm{in}},\hat{\mathbf{s}})+\sigma_{\mathrm{s,int},ii,\lambda}\Phi_{\mathrm{int},ii,\lambda}(\hat{\mathbf{s}}_{\mathrm{in}},\hat{\mathbf{s}})\right]\mathrm{d}\Omega_{\mathrm{in}} \\ & +\frac{\sigma_{\mathrm{s,int},ji,\lambda}}{4\pi}\int_{4\pi} I_{j,\lambda}(\mathbf{x},\hat{\mathbf{s}}_{\mathrm{in}})\Phi_{\mathrm{int},ji,\lambda}(\hat{\mathbf{s}}_{\mathrm{in}},\hat{\mathbf{s}})\mathrm{d}\Omega_{\mathrm{in}}, \qquad i,j=1,2 \text{ and } i\neq j\end{aligned} \tag{1}$$

where I_i represents the volume-averaged intensity of phase i; κ, σ, β, Φ, represent the absorption coefficient, scattering coefficient, and extinction coefficient, and the scattering phase function, respectively; the subscript d represents bulk properties of the phases which incorporate morphological characteristics with dimensions at the small scale ($x<1$); the subscript int represents the effective properties which incorporate morphological characteristics with dimensions at the middle scale ($x>1$). The coupled RTEs can be solved using path-length based Monte Carlo methods with appropriate large-scale boundary conditions [5].

The effective properties, σ_{ii}, σ_{ij}, β_i, Φ_{ii}, Φ_{ij}, are postulated when deriving the volume-averaged RTEs [9] and a collision-based Monte Carlo method is used for the determination of the effective properties [4]. The exact morphology is incorporated by using a level-set approach for the discretization and X-ray tomography for obtaining the 3D morphology [10].

The bulk properties account for morphological characteristics at the sub-micron scale ($x<1$) and are determined by solving the Maxwell's equations [1]. The Maxwell's equations are solved by using the discrete dipole approximation (DDA) where the complex structures are approximated by a large numbers of dipoles and the calculated dipole moments are used for the determination of the scattering and extinction efficiencies, and scattering phase function [2, 3].

3. Application to snow with soot impurities

The methodology introduced is applied to snow layers with soot impurities. The importance for the characterization of the radiative behaviour of snow is related to the fact that the total net anthropogenic radiative forcing is predicted to be 1.6 +0.8/-1.0 W/m^2, of which in average 0.1 W/m^2 are associated to changes in surface albedo due to soot impurities in snow but with a large uncertainty of +/-0.1 W/m^2 [7]. Snow and agglomerated soot have both very complex and stochastic morphologies as shown and characterized in Table 1. We choose two distinct snow samples: *i*) metamorphosed snow (mI), and *ii*) wet snow (ws), which have been incorporated using X-ray tomography [5]; and two types of soot agglomerates, each composed of 400 primary particles of 10 nm diameter and randomly accumulated [8]. Additionally, the soot agglomerates have been approximated by an equivalent sphere with the same volume as the agglomerates, and by 400 single particles with the primary particle diameter. A volume fraction of soot in the ice of 10^{-6} was exemplary chosen. The spectral complex refractive index of ice and soot has been extracted from literature [12, 11].

Table 1. Morphological characteristics and refractive index of snow and soot agglomerates.

	Morphology	Porosity	Specific surface	Particle size	Anisotropy	Fractal dimensionality	Refractive index
Metamorphosed snow (mI)	 Edge length 4mm	0.85	6 mm^{-1}	80 μm [5]	0.157	-	Ice [12]
Wet snow (ws)	 Edge length 4mm	0.38	3 mm^{-1}	666 μm [5]	0.035	-	Ice [12]
Agglomerate 1 (a1)	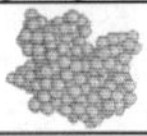	0.85	89 $μm^{-1}$	10 nm	0.141	2.412	Soot [11]
Agglomerate 2 (a2)		0.95	31 $μm^{-1}$	10 nm	0.514	2.175	Soot [11]

The spectral scattering and extinction efficiencies and scattering phase function of the soot agglomerates have been calculated using DDSCAT [2, 3] or BHMIE [1]. Together with the spectral effective scattering and extinction coefficients and scattering phase functions, these properties are used to solve eqs. (1). The calculated hemispherical-hemispherical reflectivity of an infinite mI and ws 4cm-thick snow slab diffusely irradiated is depicted in Figure 1. The difference in reflectance was up to 275% for wavelengths below 1.5 μm solely based on the middle scale morphology. For the same mid-scale morphology, variations in small scale morphologies lead to differences of up to 18%.

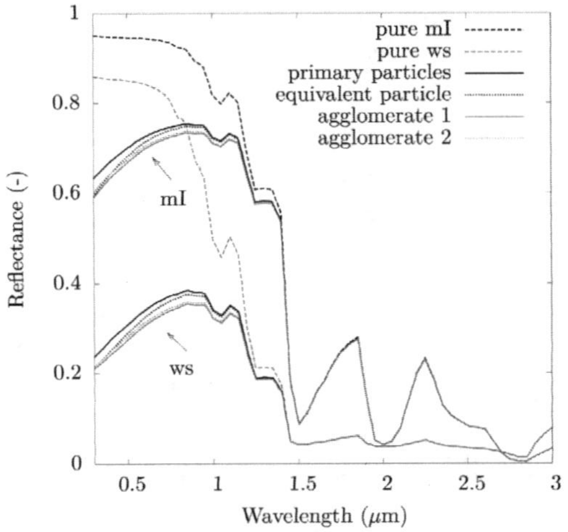

Figure 1. Reflectance of a diffusely irradiated 4 cm-thick mI and ws snow layer without (dashed) and with soot impurities modeled as *i)* uniformly distributed single soot primary particles (black solid), *ii)* uniformly distributed agglomerates modeled by an equivalent large sphere (black dashed), and *iii)* uniformly distributed agglomerates with high (grey solid) and low (grey dotted) fractional dimensionality.

References

[1] C Bohren and D Huffman 2004 John Wiley and Sons Inc., New York.
[2] BT Draine and PJ Flatau 1994 *J. Opt. Soc. Am. A*, **11** 1491–99.
[3] PJ Flatau and BT Draine 2012 *Opt. Express* **20** 1247–52.
[4] S Haussener, W Lipinski, J Petrasch, P Wyss and A Steinfeld 2009 *J.Heat Transfer* **131** 072701.
[5] S Haussener, M Gergely, M Schneebeli and A Steinfeld 2012 *J. Geophys. Res.*, **117** F03009.
[6] S Haussener and A Steinfeld 2012 *Materials* **5** 192–209.
[7] IPCC 2007 Cambridge University Press.
[8] Ü Köylü and G Faeth 1993 *Journal of Heat Transfer* **115** 409–17.
[9] W Lipinski, D Keene, S Haussener and J Petrasch 2010 *J. Quant. Spec. Rad. Trans.* **111** 2474.
[10] J Petrasch, P Wyss and A Steinfeld 2007 *J. Quant. Spec. Rad. Trans.* **105** 180–97.
[11] EP Shettle and RW Fenn 1979 *Optical Physics Division.*
[12] S Warren and E Brandt 2008 *J. Geophys. Res.* **113** 10.1029/2007JD009744.

Determination of space-dependent radiative properties in diffuse optical tomography using a wavelet multi-scale method

F. Dubot[1], Y. Favennec[2], B. Rousseau[2] and D. R. Rousse[1]

[1] Chaire t3e, École de technologie supérieure, Montréal, Canada
[2] LTN UMR CNRS 6607, Université de Nantes, France

E-mail: fabien@t3e.info

Abstract. This paper deals with the estimation of radiative property distributions of participating media from a set of light sources and sensors located on the boundaries of a medium. This is the so-called diffuse optical tomography problem. Such a non-linear ill-posed inverse problem is solved through the minimization of a cost function which depends on the discrepancy, in a least-square sense, between some measurements and associated predictions. In the present case, predictions are based on the diffuse approximation model in the frequency domain while the optimization problem is solved by the L-BFGS algorithm. To cope with the local convergence property of the optimizer and the presence of numerous local minima in the cost function, a wavelet multi-scale method associated with the L-BFGS method is designed.

1. Introduction

The diffuse optical tomography (DOT) problem is an inverse problem in which the spatial distributions of absorption and scattering coefficients of a participating medium are sought from a set of light sources and sensors located on the frontier ($\partial\Omega$) of the probed medium (Ω). Such a non-linear ill-posed inverse problem is solved through the minimization of a cost function which depends on the discrepancy, in a least-square sense, between some measurements and associated predictions. Among employed methods to solve this problem, the Limited-memory Broyden-Fletcher-Goldfarb-Shanno (L-BFGS) method has proven its worth [1, 2]. However, some problems remain to be overcome such as the stability with respect to the initial guesses and the blurring effect of the reconstructed images due to the need for relatively strong regularization tools. In recent years, the wavelet multi-scale method (WMM) has emerged in the field of inversion [3]. This method relies on a reformulation of the original inverse problem into a sequence of sub-inverse problems of different scales using wavelet transform, from the largest scale to the smallest one. It has been shown that the WMM can enhance stability of inversion, accelerate convergence and cope with the presence of local minima to reach the global minimum. The rationale behind this success is that the cost function shows stronger convexity and has less minima at larger scale such that the global minimum can be achieved. Best results can thus be obtained, using this solution to the initialization of the optimization problem at the lower scale until finding the minimum at the original scale. This paper is dedicated to the application of a WMM associated with the L-BFGS method to the specific inverse problem of the DOT and aims to improve the efficiency of the ordinary L-BFGS algorithm.

2. Inverse model

The forward model used in this study is the bidimensional diffuse approximation (DA) model in the frequency domain [4]. This equation leads to compute the photon density $\varphi : \Omega \mapsto \mathbb{C}$ that depends on space-dependent absorption (κ) and reduced scattering (σ') coefficients, and which values at the nodes of the sensors will be called predictions. Let $\partial\Omega_s = \partial\Omega_s^1 \cup \ldots \partial\Omega_s^L$ be a set of L surfacic sensors on $\partial\Omega$ and let us consider K collimated sources at different locations along $\partial\Omega$. For each source, let $\breve{\varphi}_i : \Omega \mapsto \mathbb{C}$ be equal to measurements of the transmitted photon density. Predictions φ_i and measurements $\breve{\varphi}_i$ are integrated to the cost function to be minimized:

$$j(\kappa, \sigma') = \frac{1}{2} \sum_{i=1}^{K} \int_{\partial\Omega_s} \left| \frac{\varphi_i - \breve{\varphi}_i}{\breve{\varphi}_i} \right|^2 d\zeta \tag{1}$$

Because of the different orders of magnitude between the two radiative properties κ and σ', a strategy should be developed to facilitate the simultaneous reconstruction of these properties if a gradient-based method is used to perform this minimization. Indeed, the cost function gradient parts associated with both radiative properties also differ by roughly the same order of magnitude, which is very bad for the convergence. In order to speed-up the iterative convergence to the local minimum, the proposed strategy consists in performing a scaling on the parameters at the beginning of the optimization problem. Choosing an a priori function for both radiative properties, say κ_{ap} and σ'_{ap}, one searches parameters that fluctuate about unity. This adimensionalization leads to recover both $\varkappa(\mathbf{x}) = \kappa(\mathbf{x})/\kappa_{ap}(\mathbf{x})$ and $\varsigma(\mathbf{x}) = \sigma'(\mathbf{x})/\sigma'_{ap}(\mathbf{x})$ for which magnitude is of order one approximately for both coefficients. The inverse problem is then formulated as a constrained optimization problem of the form:

$$\text{Find } (\bar{\varkappa}, \bar{\varsigma}) = \arg\min_{\varkappa,\varsigma} j(\varkappa, \varsigma) \text{ subject to } \varphi_i \text{ solution of the DA model, } \forall i \tag{2}$$

The main tool used for the development of the WMM is the projection onto wavelet bases which relies on Mallat's forward and inverse fast wavelet transforms (FWT) [6]. Let us consider a discrete signal a_N at scale N, i.e. whose length is 2^{-N}, $N \leq 0$. The forward FWT computes the vectors of scale and detail coefficients at scale $N+1$ of the signal a_N, a_{N+1} and b_{N+1} of length 2^{-N-1}, with the help of discrete convolutions between a_N and conjugate mirror filters h and g, respectively. The inverse FWT allows to reconstruct the original signal a_N from discrete convolutions between a_{N+1} and h, and between b_{N+1} and g. The projection of the signal a_N, for example from scale N to scale $N+2$, consists in the following operations: (i) compute a_{N+2} using the forward FWT with a_N then a_{N+1}; (ii) reconstruct a_N using the inverse FWT with a_{N+2} then a_{N+1}, assuming the vectors of detail coefficients b_{N+2} and b_{N+1} are equal to 0. Note that only the filter h, which fully determines the wavelet basis and may be interpreted as a low pass discrete filter which smooths the data, is used when projecting the signal.

The L-BFGS method [5] associated to a quadratic line-search method is used to solve the optimization problem (2). For CPU time considerations, cost function gradients are computed by the very fast adjoint-state method from the state (φ_i) and adjoint (φ_i^*) variables. For the sake of simplicity, the WMM is presented with the domain Ω discretized on a grid of $2^{-N} \times 2^{-N}$ points, $N \leq 0$, with each sensor constituted of 2^{-M} pointwise measurements on the grid points, $N \leq M \leq 0$, and with all functions involved in the inverse method discretized by P1-finite elements. Thus, the vectors (images) of scale coefficients at the smallest scale associated to the predictions and measurements on sensors (state and adjoint variables) can be simply built from the values of these functions on the grid points. The WMM consists in applying the L-BFGS method until convergence, from the largest scale ($N - M$) to the smallest one (N), with, at a given scale, initializing the radiative properties to the solution of the optimizer at the upper scale. Ultimately, this method is essentially based on careful computations of the cost function and cost function gradient at different scales using the projection operator.

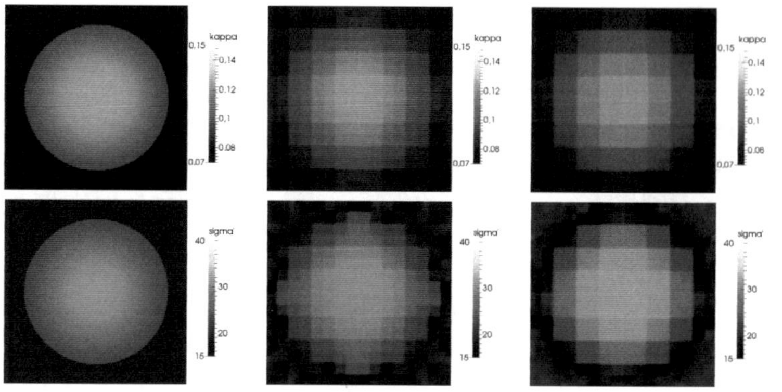

Figure 1. Reconstructions with the Haar wavelet. Top, results for κ; bottom, results for σ'. From left to right: target properties; $\kappa_{ap} = 0.08$, $\sigma'_{ap} = 20$; $\kappa_{ap} = 0.08$, $\sigma'_{ap} = 22.5$.

3. Numerical results

A selected numerical result is presented in this section to assess the validity of the proposed method. The domain, Ω, is a 4 cm square. The regular grid associated to Ω is composed of 128×128 points ($N = -7$). On each side of the square, one source and two sensors are located. Each sensor contains 16 pointwise photon density measurements ($M = -4$). Synthetic data are considered for this numerical test. These data, $\breve{\varphi}$, representing the pseudo-experimental measurements, are built on a finer mesh (255×255 points) than that of generating the predictions φ in order to avoid the inverse crime. Then, a multiplicative white gaussian noise is applied to $\breve{\varphi}$ at the nodes of the sensors to simulate the noise inherent to experimental devices, i.e. $\breve{\varphi}_{\text{noisy}} = \breve{\varphi}(1 + 0.01 \times m)$ where $m \sim \mathcal{N}(0, 1)$. The FWT is applied with the Haar wavelet ($h = [2^{-1/2}, 2^{-1/2}]$) to project the synthetic data at scales $-3, \ldots, 0$ and thus, to obtain a series of inverse problems for scales $-7, \ldots, -3$. Target properties are given in Figures 1a and 1d. Simultaneous reconstructions of κ and σ' with a priori functions $(\kappa_{ap}, \sigma'_{ap}) = (0.08, 20)$ and $(\kappa_{ap}, \sigma'_{ap}) = (0.08, 22.5)$ were investigated with the L-BFGS method and the WMM using the Haar wavelet. It was observed that, for these specific initial guesses, the L-BFGS method failed to converge unlike the WMM which provided quite accurate results (see Fig. 1). Moreover, for other initial guesses where the two methods converged, it is found that the WMM is faster and provides more accurate reconstructions than the L-BFGS method. For example, with $(\kappa_{ap}, \sigma'_{ap}) = (0.09, 22.5)$, the L-BFGS method converged in 1 049 s while the WMM converged in 235 s (Intel i3 2.4 GHz, RAM 6 GB). Although the per-iteration CPU times are slightly longer for the WMM, fewer iterations are required to reach the convergence. Finally, to avoid the "square-shaped" aspect of the obtained reconstructions due to the use of the Haar wavelet, other wavelets were considered. However, because an edge problem appears when considering the FWT with another wavelet than the Haar wavelet, better results are not obtained. An extension of the proposed WMM with the wavelet extrapolation method [7], which does not produce undesirable edge effects, is in progress to obtain an efficient WMM using other wavelets.

References

[1] Klose A D and Hielscher A H 2003 *Inverse Problems* **19** 387
[2] Balima O, Boulanger J, Charette A and Marceau D 2011 *J. Quant. Spectrosc. Radiat. Transfer* **112** 1235–1240
[3] Lei J, Liu S, Li Z, Sun M and Wang X 2011 *Applied Mathematical Modelling* **35** 2585–2606
[4] Arridge S R 1999 *Inverse problems* **15** R41
[5] Liu D and Nocedal J 1989 *Mathematical Programming: Series A and B* **45** 503–528
[6] Mallat S 1999 *A wavelet tour of signal processing* (Academic press)
[7] Williams J R and Amaratunga K 1997 *Journal of Fourier Analysis and Applications* **3** 435–449

Journal of Physics: Conference Series **676** (2016) 012005 doi:10.1088/1742-6596/676/1/012005

kspectrum: an open-source code for high-resolution molecular absorption spectra production

V. Eymet, C. Coustet, B. Piaud
Méso-Star SAS, 8 rue des pêchers, 31410 Longages, France
E-mail: vincent.eymet@meso-star.com

Abstract. We present the *kspectrum* scientific code that produces high-resolution synthetic absorption spectra from public molecular transition parameters databases. This code was originally required by the atmospheric and astrophysics communities, and its evolution is now driven by new scientific projects among the user community. Since it was designed without any optimization that would be specific to any particular application field, its use could also be extended to other domains. *kspectrum* produces spectral data that can subsequently be used either for high-resolution radiative transfer simulations, or for producing statistic spectral model parameters using additional tools. This is a open project that aims at providing an up-to-date tool that takes advantage of modern computational hardware and recent parallelization libraries. It is currently provided by *Méso-Star* (http://www.meso-star.com) under the CeCILL license, and benefits from regular updates and improvements.

1. Introduction

The production of synthetic high-resolution molecular spectra is a central concern in many scientific domains involving radiative transfer: in atmospheric sciences, the inversion of high-resolution spectral signals measured from space is used in order to retrieve composition and temperature profiles. In astrophysics, the measurements of emitted radiation can provide insightful information about the composition of nebulae and stellar atmospheres. Recently, astrobiologists have been interested in retrieving atmospheric composition and ground temperatures on distant exoplanets from radiation that is both emitted and reflected by the planet.

In study domains that involve high temperatures, radiation solvers both rely on high-resolution spectra [1] and statistical models [2] such as the Malkmus transmittivity model [3] that has historically been heavily used. Data sets that can be used in order to obtain Malkmus' model parameters $\bar{k}$ and ϕ (respectively, the average absorption coefficient and the overlap parameters within a given narrowband) for a wide range of thermodynamic conditions (pressure, temperature and concentrations) for the main radiatively active species in flames (H_2O, CO_2 and CO) have been derived from measurements [4, 5]. This work is more than ever relevant [6] in the context of new theoretical developments on average transmittivity models that can also take into account the spatial heterogeneities of the participating medium [7]. The validation of radiative transfer results obtained with such spectral models must be performed using reference results, obtained with high-resolution spectra. In more recent approaches, it has been shown that both the representation of heterogeneities and the spectral integration could be achieved using null-collision algorithms [8, 9] that require using a modified absorption coefficient field

$\hat{k}$ that should be as close as possible from an upper-bound of the real absorption coefficient field within the medium; this modified absorption coefficient field has been estimated using high-resolution spectra until now, but this work can certainly also take advantage of statistical spectral models approaches.

Building synthetic spectra is a computationally heavy task, since transition databases keep growing, both in the number of molecular species that are taken into account and in the number of transitions that are identified for each species: transition databases originally gathered information mainly about H_2O and CO_2. Within the HITRAN database, the number of molecular species grew from 39 in the 2004 version [10] to 42 in the 2008 version [11], and the 2012 version [12] counts 47 species. Ambitious projects such as ExoMol [13] are specialized in gathering transition parameters for many complex molecules that are found in planetary and stellar atmospheres.

The *kspectrum* project was initiated by the community of climate study, during a modelization effort on the atmosphere of Venus [14]: the spectral data used for the gas mixture, while the best available at that time, had hardly been referenced and it was not possible to produce it for different temperature or composition profiles. Since then, the evolution of the code was driven by the needs of its users [15, 16] and the various scientific projects it has been involved in. The code has been used in climate modeling of both solar-system planets [17, 18] and exoplanets [19, 20, 21]. Recently, it was also used in order to better define the boundaries of the habitability zone around main-sequence stars [22]. And a new study of Venus atmosphere that should be published in the next weeks [23] also uses high-resolution spectra produced by *kspectrum*.

kspectrum is an open-source project, aiming at giving a complete control and traceability about spectral data production, and that is progressively being enriched by the experience acquired while interacting with the users. The code is distributed under the CeCILL license and maintained by Méso-Star (`http://www.meso-star.com/en_Products.html`). The main purpose of *kspectrum* is to provide a framework that can be easily modified and adapted to specific needs. The code has been designed in order to be flexible in terms of architecture and reliable, and can easily be adapted to various computational hardware and new parallelization libraries.

When computing atmospheric absorption spectra, an adapted line profile truncation is commonly used. Then, an ad-hoc continuum is added to the resulting spectra. This continuum accounts for both the underestimation of the absorption coefficient because of the truncation, and for specific additional opacity (namely, collision-induced absorption, CIA) that is specific to the study case. This results in very well validated codes. In *kspectrum*, we privileged versatility: even though the code can perform some CIA computation, it is preferred not to use any additional continuum in molecular absorption spectra computations. Then, the user has the possibility to add its own continuum in order to account for specific opacity that has not been taken into account (and possibly to correct truncation effects). Since nothing has been hard-coded for specific applications, the code can be used in a wide variety of cases and scientific fields.

2. Description of the code

The main purpose of *kspectrum* is to compute high-resolution molecular absorption spectra for a wide range of thermodynamic conditions, using the 47 molecular species referenced in the 2012 version of the HITRAN database [12]. However, the code is not restricted to using the HITRAN database, and the user can control whether or not high-temperature data is used when it is available for these molecular species.

2.1. Physics

The *kspectrum* code has been designed in order to address a wide variety of applications. It can handle modern public transition databases: HITRAN [10, 11, 12], HITEMP [24] and CDSD

[25, 26] ; high-temperature versions (HITEMP-2010 for H_2O and CDSD-4000 for CO_2) are supported as well. In terms of line profiles, the user can choose between using a plain Lorentz profile and the Voigt line profile, and several sub-lorentzian corrections have been included for the CO_2 molecule in the 1.18 μm and 2.7 μm bands [27], in the 2.3 μm band [28] and the 4.3 μm band [29] (see figure **??**). Some collision-induced absorption continuum parameterization has also been included [30, 31], but CIA continuum will preferably be computed separately from molecular absorption. Finally, the code can handle a separate abundance for each isotopologue of each molecular species.

2.2. Computational aspects

kspectrum was originally designed from the project needs with two goals in mind:

(i) to ensure a given uncertainty over each value of k(ν) that is computed

(ii) to ensure a given uncertainty when using a linear interpolation between two consecutive values of k(ν)

The first point (i) imposes to be able to take into account the contribution of each known transition to the total absorption coefficient at a given wavenumber ν, even lines whose center is defined at a (possibly) infinite distance of the considered value of ν. The fact that the line profile is, in practice, poorly known after a few tens of inverse centimeters from the line center, should not be a constraint at this point: being able to compute synthetic spectra with and without truncation at least provides some insights about the boundaries where the "real" (physical) spectrum lies; this point is further developed in section 3. The second point (ii) imposes that the spectral step (between two consecutive values of ν) has to be computed in order to fulfill this condition. The method used is conceptually very similar to the idea of multiple spectral grids developed in [32].

In terms of computation time, synthetic spectra computed with such a full control of the numerical uncertainty are possible, but difficult to achieve. The code was subsequently modified in order to enable faster computations. One major source of computation time gain is the spectral discretization: in addition to the reference spectral discretization, other algorithms have been implemented, with decreasing computational requirements:

- using a given number of discrete wavenumbers for representing line centers and inter-line regions
- using only line centers as the spectral grid
- constant spectral step
- pre-computed spectral grid

Additionally, the user has the possibility to remove least intense lines for the spectral discretization step. This feature is particularly useful when using high-temperature transition databases that include several hundred millions of lines. Of course, the spectral grid obtained with one of these faster spectral discretization algorithms or when using a limited number of transitions does no longer satisfies condition (ii).

Further computational gains have been found by the implementation of:

- Line profile truncation, either at a fixed distance from the line center, or at a fixed multiple of the line width that differs from one line to the other. As illustrated in figure 2, the effects of truncating line profiles can result in a under-estimation of the absorption coefficient, and possibly the presence of unrealistic steps in the spectrum.
- Computing each value of k(ν) using a convergent approach: the contribution of most intense lines is computed first. The expected convergence of the sum of line contributions as line intensities decrease is periodically checked, and the algorithm stops computing line contributions when a convergence criterion is satisfied.

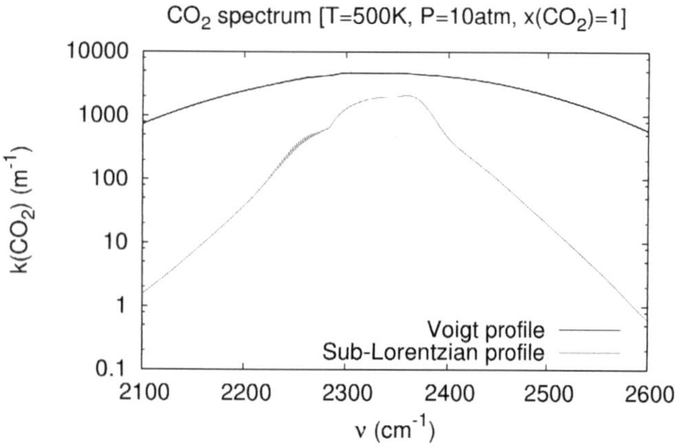

(a)

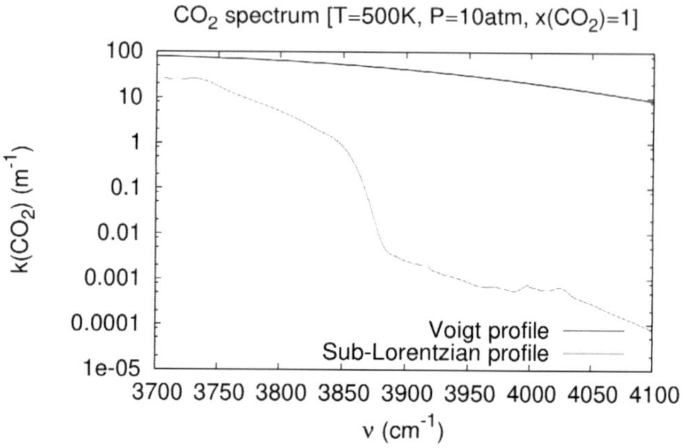

(b)

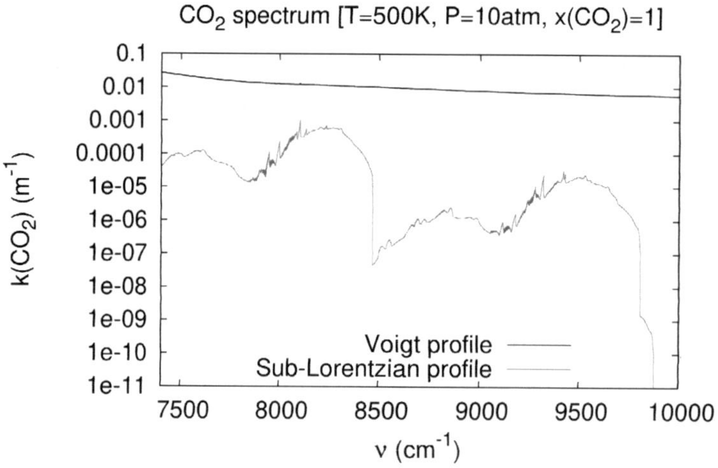

(c)

Figure 1. Synthetic spectra of CO_2 at a temperature of 500K and a pressure of 10atm, computed with and without sub-lorentzian line profile correction in the (a) 4.3 μm, (b) 2.7 μm and (c) 1.18 μm windows.

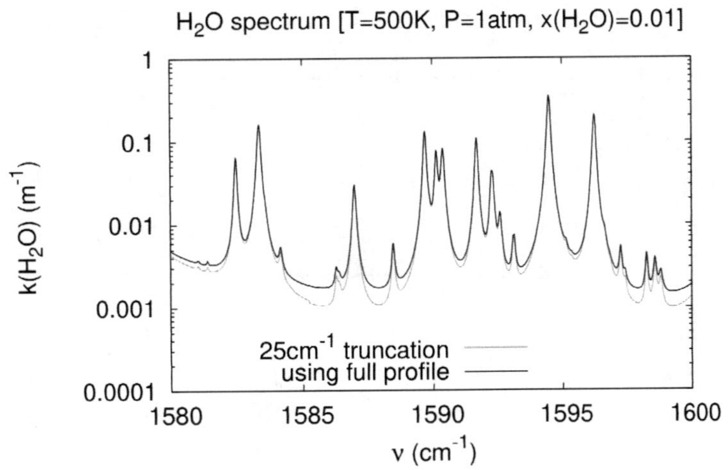

Figure 2. Synthetic spectrum of H_2O at a temperature of 500K and a pressure of 1 atm, computed using the full Voigt profile at a infinite distance from line centers, compared with the same spectrum obtained with a truncation of the line profile at 25 cm^{-1} from the line centers.

The consequence of using either of these features is that the accuracy over each value of $k(\nu)$ is no longer guaranteed (see condition (i)).

When possible, the various stages involved in spectra computation have been parallelized in *kspectrum*. These stages include:

- tabulation of the Voigt function, required by the reference spectral discretization algorithm that must satisfy condition (ii).
- the identification of transition parameters location in each LBL file that should be used, for a number of pre-defined spectral intervals. This step is required for a faster subsequent access to this data.
- the spectral discretization of any spectral interval.
- finally, the computation of $k(\nu)$ values.

The computational load of each process is balanced using a very basic but efficient algorithm: the total computational load (for instance, computing a set of N_k values of $k(\nu)$ once the spectral discretization has been completed) is divided into N_{task} tasks (with N_{task} much greater than the number of processes). Each process will then be allocated one task at a time. In this example, the elementary task involves the computation of $\frac{N_k}{N_{task}}$ values of $k(\nu)$. Once the task is over, the process sends back its results to the master process, and is allocated a new task, until all tasks have been performed. This rudimentary heuristic ensures that each process is near-to optimally used: the number of computing processes is generally equal to the number of cores available. If, for any reason, one or several cores are required by system processes or other codes, *kspectrum* processes that are running on these cores will naturally manage fewer tasks than processes that run on fully available cores. This results in globally faster computations than if the number of tasks is equal to the number of computing processes.

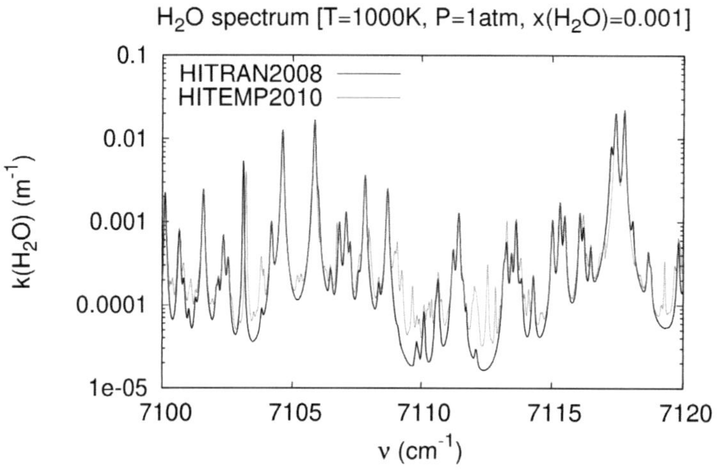

Figure 3. Synthetic spectra of H_2O at a temperature of 1000K and a pressure of 1 atm, computed using both the 2008 version of HITRAN and the 2010 version of HITEMP.

Handling the latest transition databases has also been a challenge: the HITEMP-2010 database includes 114 millions of transitions for H_2O, for a total of 18 Gb. Even if all parameters are not required for spectra computations, it is not yet possible to load all the required information in RAM at once on a mainstream computer. This is even worse for CDSD-4000, that holds 73 Gb of data over 2768 files for 626 millions of CO_2 transitions. When not using the convergence algorithm (computing first the contribution of most intense transitions), the contribution of every line has to be added for every value of $k(\nu)$. Since loading all the required data at once is not possible, *kspectrum* uses a different strategy: it loads only the parameters for a limited number of transitions, and computes their contributions to the required value of $k(\nu)$. This procedure is repeated until the parameters of all transitions have been read and all contributions have been computed. This multi-step approach results in a slower computation, but one major advantage in terms of flexibility is that the code can run with very limited RAM capacity. The lower is the RAM capacity, the greater will be the computation time for a given spectrum because of a greater number of successive steps. Figure 3 illustrates the impact of using a high-temperature transition database on a H_2O spectrum.

Finally, *kspectrum* performs frequent backups of the current state of the computation (in both spectral discretization and actual spectrum computation stages), so that the computation can be resumed from the latest backup in the event of a power outage or a maintenance reboot. This feature may seem anecdotic, but has always been very appreciated by users.

3. Academic test case

As far as we know, using a Lorentz or a Voigt profile at an infinite distance from line centers is not correct, because in practice line profiles are not well known after a few tens or hundreds of inverse centimeters. So taking into account the contribution of every transition, even at an infinite distance from the considered wavenumber, results in an overestimation of the spectrum. Similarly, truncating every line profile at a fixed distance of 25 cm^{-1} results in a underestimation

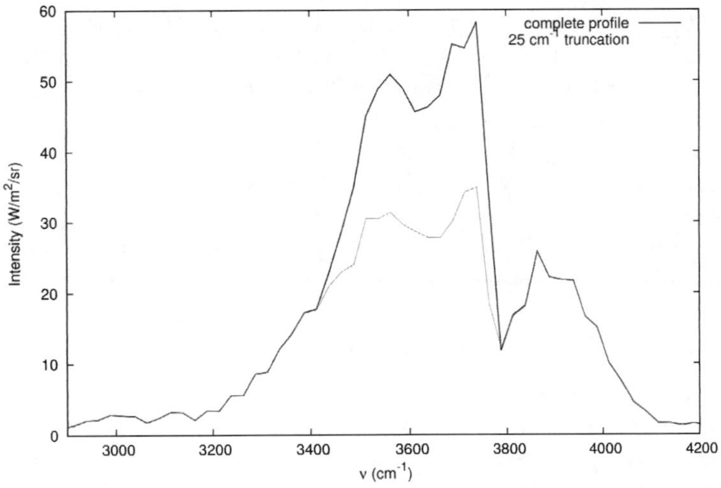

Figure 4. Specific intensity emitted by a one-meter homogeneous gas column containing H_2O, CO_2 and CO (all three species at a 0.1 molar fraction), at 1000K and 1atm. High-resolution results have been averaged over 25 cm^{-1} spectral intervals. The black line has been obtained using a full line profile and taking into account the contribution of every known transition, and the red line shows the same result, obtained using a 25 cm^{-1} line profile truncation and rejecting least intense lines.

of the spectrum. Without speaking of additional opacity sources such as collision-induced absorption, the real absorption spectrum lies somewhere between these two limits.

Figure 4 illustrates the kind of tests that can be performed with *kspectrum*: two synthetic spectra have been computed for a 1000K, 1atm participating media, containing H_2O, CO_2 and CO with equal molar fractions of 0.1. Both spectra have been obtained from the 2012 version of the HITRAN database. The first spectrum (black line) was computed using a full line profile, and taking into account every known transition in the database for these species. The second spectrum (red line) was obtained on the basis of a 25 cm^{-1} truncation of the line profile, and using the convergence algorithm for neglecting least intense transitions. Then the specific intensity emitted by a one-meter homogeneous gas column was computed over the high-resolution spectral grid, and has been averaged over 25 cm^{-1} spectral intervals in order to obtain the results displayed in figure 4. Off course, the emphasis has been put on a spectral region where the difference between the two results is significant: in most parts of the infrared region, the two spectra match perfectly. The overall effect can be quantified by a comparison of the total emitted specific intensity, integrated in the [150;9325] cm^{-1} longwave region: it is 4564.47 $W/m^2/sr$ using a full line profile and 4366.91 $W/m^2/sr$ when a truncation is applied; the effect of using a truncation in this example is at most 5% over the integrated specific intensity, even if it can locally be responsible of much more important differences in specific spectral regions.

4. Conclusion

kspectrum has been progressively improved over the last few years. It has already been used for a number of scientific projects in the atmosphere and astrophysics communities: climate modeling of Saturn's atmosphere [18] and of exoplanets [21, 15], confirming hypothesis about exoplanets observed by space telescopes [20], simulate the climatic evolution of primitive Earth-like atmospheres [19] and better defining the boundaries of habitable zones [22]. The code has been thought to be robust and reliable, and can run as well on cheap desktop computers as on dedicated servers. Users have been interested in both the possibility to compute absorption spectra without compromises, and the fact that nothing specific to a given application has been hard-coded, making *kspectrum* a very versatile tool.

In terms of perspective, there is already a clearly identified need to make *kspectrum* compatible with transitional data available in the ExoMol database (`http://www.exomol.com/`). This would be very helpful for the astrophysics and astrobiology communities. This would imply some development time, mainly because of the large number of molecular species involved and the numerous formats used to provide various transition parameters within this database.

kspectrum results can be used directly as input for radiative transfer codes, or may be used by the following companion codes:

- *kdistribution*: does exactly what its name says: to produce k-distribution data sets from *kspectrum* high-resolution spectra, for later use in radiative transfer solvers that are using the correlated-k spectral model [33]. Using directly high-resolution spectra within radiative transfer solvers is also very useful; however, since the default behaviour of *kspectrum* is to produce unstructured spectral grids, the radiative transfer solver has to be able to interpolate absorption coefficients from its input spectra over the spectral grid used for radiative transfer computations.
- *gas_mix*: this is a code that produces high-resolution spectrum for any gas mixture, in any thermodynamic conditions (just like what *kspectrum* does, but much faster) using a database of pre-computed high-resolution spectra over a given (P/T/x) grid, for a number of molecular species. This code is able to take into account new spectra, as they are added to the database: the (P/T/x) grid over which spectra are computed does not need to be structured, and can be progressively improved. This idea has been initiated some years ago within the Starwest group and led to the development of the "High Resolution Spectra" database (`http://www.starwest.ups-tlse.fr/vincent-eymet/high-resolution-spectra/`). This first database was computed for a limited number of pressure levels and molecular species, which restricted its use mainly to combustion applications, and is now outdated in the sense that it was computed using the HITRAN-2008 database. But the main idea behind this project is still valid, and could eventually result in a online tool for providing fast access to spectral data for radiative transfer applications, similarly to what can be obtained using `http://spectra.iao.ru/`

References

[1] R. Tao and F.M. Modest. A hybrid wavenumber selection scheme for line-by-line photon monte carlo simulations in high-temperature gases. *Journal of Heat Transfer*, 135(8), 2013.

[2] Gopalendu P. and Modest M.F. A narrow band-based multiscale multigroup full-spectrum k-distribution method for radiative transfer in nonhomogeneous gas-soot mixtures. *Journal of Heat Transfer*, 132(2), 2009.

[3] W. Malkmus. Random lorentz band model with exponential-tailed S^{-1} line-intensity distribution function. *Journal of Optical Society of America*, 57:323–329, 1967.

[4] J. Taine and A. Soufiani. Gas ir radiative properties: from spectroscopic data to approximate models. *Advances in Heat Transfer*, 33:295–414, 1999.

[5] A. Soufiani and J. Taine. High temperature gas radiative property parameters of statistical narrow band model for H_2O, CO_2 and CO and correlated-k model for H_2O and CO_2. *Int. J. Heat and Mass Transfer*, 40(4):987–991, 1997.

[6] Ph. Rivière and A. Soufiani. Updated band model parameters for H_2O, CO_2, CH_4 and CO radiation at high temperature. *Intervational Journal of Heat and Mass Transfer*, 55:3349–3358, 2012.

[7] André F. and Vaillon R. Generalization of the k-moment method using the maximum entropy principle. application to the nbkm and full spectrum slmb gas radiation models. *Journal of Quantitative Spectroscopy and Radiative Transfer*, 113:1508–1520, 2012.

[8] M. Galtier. Integral formulation of null-collision monte carlo algorithms. *Journal of Quantitative Spectroscopy and Radiative Transfer*, 125:57–68, 2013.

[9] M. Galtier. *Approche statistique du rayonnement dans les milieux gazeux hétérogènes: de l'échantillonnage des transitions moléculaires au calcul de grandeurs radiatives.* PhD thesis, Doctorat de l'Université de Toulouse, 2014.

[10] L.S. Rothman, D. Jacquemart, A. Barbe, D.C. Benner, M. Birk, L.R. Brown, M.R. Carleer, C.Jr. Chackerian, K. Chance, L.H. Coudert, V. Dana, V.M. Devi, J.-M. Flaud, R.R. Gamache, A. Goldman, J.-M. Hartmann, K.W. Jucks, A.G. Maki, J.-Y. Mandin, S.T. Massie, J. Orphal, A. Perrin, C.P. Rinsland, M.A.H. Smith, J. Tennyson, R.N. Tolchenov, R.A. Toth, V.J. Auwera, P. Varanasi, and G. Wagner. The HITRAN 2004 molecular spectroscopic database. *Journal of Quantitative Spectroscopy and Radiative Transfer*, 96:139–204, 2005.

[11] L.S. Rothman, L.E. Gordon, A. Barbe, D.C. Benner, P.F. Bernath, M. Birk, V. Boudon, L.R. Brown, A. Campargue, J.-P. Champion, K. Chance, L.H. Coudert, V. Dana, V.M. Devi, S. Fally, J.-M. Flaud, R.R. Gamache, A. Goldman, D. Jacquemart, L. Kleiner, N. Lacome, W.J. Lafferty, J.-Y. Mandin, S.T. Massie, S.N. Mikhailenko, C.E. Miller, N. Moazzen-Ahmadi, O.V. Naumenko, A.V. Nikitin, J. Orphal, V.I. Perevalov, A. Perrin, A. Predoi-Cross, C.P. Rinsland, M. Rotger, M. Simeckova, M.A.H. Smith, K. Sung, S.A. Tashkun, J. Tennyson, R.A. Toth, A.C. Vandaele, and V.J. Auwera. The HITRAN 2008 molecular spectroscopic database. *Journal of Quantitative Spectroscopy and Radiative Transfer*, 110:533–572, 2009.

[12] L.S. Rothman, L.E. Gordon, Y. Bakikov, A. Barbe, D.C. Benner, P.F. Bernath, M. Birk, L. Bizzocchi, V. Boudon, L.R. Brown, A. Campargue, K. Chance, L.H. Coudert, V.M. Devi, B.J. Drouin, A. Fayt, J.-M. Flaud, R.R. Gamache, J.J. Harrison, J.-M. Hartmann, C. Hill, J.T. Hodges, D. Jacquemart, A. Jolly, J. Lamouroux, R.J. Le Roy, G. Li, D.A. Long, S.T. Lyulin, O.M. Mackie, S.T. Massie, S. Mikhailenko, H.S.P. Müller, O.V. Naumenko, A.V. Nikitin, J. Orphal, V. Perevalov, A. Perrin, E.R. Polotseva, C. Richard, M.A.H. Smith, E. Starikova, K. Sung, S.A. Tashkun, J. Tennyson, G.C. Toon, Vl.G. Tyuterev, and G. Wagner. The HITRAN 2012 molecular spectroscopic database. *Journal of Quantitative Spectroscopy and Radiative Transfer*, 130:4–50, 2013.

[13] J. Tennyson and S. N. Yurchenko. Exomol: molecular line lists for exoplanet and other atmospheres. *Monthly Notices of the Royal Astronomical Society*, 425:21–33, 2012.

[14] V. Eymet, R. Fournier, J.-L. Dufresne, S. Lebonnois, F. Hourdin, and M.A. Bullock. Net-Exchange parameterization of thermal infrared radiative transfer in Venus' atmosphere. *Journal of Geophysical Research*, 114, E11008, DOI:10.1029/2008JE003276, 2009.

[15] R. Wordsworth, F. Forget, F. Selsis, J.-B. Madelaine, E. Millour, and V. Eymet. Is gliese 581d habitable? some constraints from radiative-convective climate modeling. *Astronomy and Astrophysics*, DOI: 10.1051/0004-6361/201015053, 2010.

[16] R. Wordsworth, F. Forget, F. Selsis, and V. Eymet. Infra-red collision-induced absorption in dense CO_2 atmospheres. *Icarus*, DOI: 10.1016/j.icarus.2010.06.010, 2010.

[17] B. Charnay, F. Forget, R. Wordsworth, J. Leconte, E. Millour, F. Codron, and A. Spiga. Exploring the faint young Sun problem and the possible climates of the Archean Earth with a 3-D GCM. *Journal of Geophysical Research (Atmospheres)*, 118(D17):10414, September 2013.

[18] S. Guerlet, A. Spiga, M. Sylvestre, M. Indurain, T. Fouchet, J. Leconte, E. Millour, R. Wordsworth, M. Capderou, B. Bézard, and F. Forget. Global climate modeling of Saturn's atmosphere. Part I: Evaluation of the radiative transfer model. *Icarus*, 238:110–124, August 2014.

[19] J. Leconte, F. Forget, B. Charnay, R. Wordsworth, and A. Pottier. Increased insolation threshold for runaway greenhouse processes on Earth-like planets. *Nature*, 504:268–271, December 2013.

[20] B. Samuel, J. Leconte, D. Rouan, F. Forget, A. Léger, and J. Schneider. Constraining physics of very hot super-Earths with the James Webb Telescope. The case of CoRot-7b. *Astronomy and Astrophysics*, 563:A103, March 2014.

[21] J. Leconte, F. Forget, B. Charnay, R. Wordsworth, F. Selsis, E. Millour, and A. Spiga. 3D climate modeling of close-in land planets: Circulation patterns, climate moist bistability, and habitability. *Astronomy and Astrophysics*, 554:A69, June 2013.

[22] R. Kopparapu, R. Ramirez, J.F. Kasting, V. Eymet, T.D. Robinson, S. Mahadevan, R.C. Terrien, S. Domagal-Goldman, V. Meadows, and R. Deshpande. Habitable zones around main-sequence stards: new estimates. *Astrophysical Journal*, 765, issue 2, 2013.

[23] S. Lebonnois, V. Eymet, C. Lee, and J.V. d'Ollone. Analysis of the radiative budget of venus atmosphere based on infrared net exchange rate formalism. *To be submitted to Earth and Space science*, 2015.

[24] L.S. Rothman, I.E. Gordon, R.J. Barber, H. Dothe, R.R. Gamache, A. Goldman, V.I. Perevalov, S.A.

Tashkun, and J. Tennyson. HITEMP, the high-temperature molecular spectroscopic database. *Journal of Quantitative Spectroscopy and Radiative Transfer*, 111:2139–2150, 2010.

[25] S.A. Tashkun, V.I. Perevalov, J.-L. Teffo, A.D. Byko, and N.N. Lavrentieva. CDSD-1000, the high-temperature carbon dioxide spectroscopic databand. *Journal of Quantitative Spectroscopy and Radiative Transfer*, 82:165–196, 2003.

[26] S.A. Tashkun and V.I. Perevalov. CDSD-4000, High-resolution, high-temperature carbon dioxide spectroscopic databank. *Journal of Quantitative Spectroscopy and Radiative Transfer*, 112, issue9:1403–1410, 2011.

[27] J.B. Pollack, J.B. Dalton, D.H. Grinspoon, R.B. Wattson, R. Freedman, D. Crisp, D.A. Allen, B. Bezard, C. DeBergh, L.P. Giver, Q. Ma, and R. Tipping. Near-infrared light from Venus' nightside: A spectroscopic analysis. *Icarus*, 103:1–42, 1993.

[28] M.V. Tonkov, N.N. Filipov, V.V. Bertsev, J.P. Bouanich, Nguyen Van-Thanh, C. Brodbeck, J.M. Hartmann, C. Boulet, F. Thibault, and R. LeDoucen. Measurements and empirical modeling of pure CO_2 absorption in the 2.3-μm region at room temperature: far wings, allowed and collision-induced bands. *Applied Optics*, 35 No. 24:4863–4870, 1996.

[29] M.Y. Perrin and J.M. Hartmann. Temperature-dependant measurements and modeling of absorption by $CO_2 - N_2$ mixtures in the far line-wings of the $4.3\mu m$ CO_2 band. *Journal of Quantitative Spectroscopy and Radiative Transfer*, 42:311–317, 1989.

[30] M. Gruszka and A. Borysow. Computer simulation of the far infrared collision induced absorption spectra of gaseous CO_2. *Molecular physics*, 93 No. 6:1007–1016, 1998.

[31] Y. Baranov, W.J. Lafferty, and G.T. Fraser. Infrared spectrum of the continuum and dimer absorption in the vicinity of the O_2 vibrational fundamental in O_2/CO_2 mixtures. *Journal of Molecular Spectroscopy*, 228:432–440, 2004.

[32] C. Caliot. *Modélisation et simulation de l'émission énergétique et spectrale d'un jet réactif composé de gaz et de particules à haute température issus de la combustion d'un objet pyrotechnique.* PhD thesis, Thèse de doctorat, Insitut National Polytechnique de Toulouse, 2006.

[33] A. A. Lacis and V. Oinas. Description of the correlated-k distribution method for modeling nongray gaseous absorption, thermal emission, and multiple scattering in vertically inhomogeneous atmospheres. *Journal of Geophysical Research*, 96:9027–9063, 1991.

Some speed-up strategies for solving inverse radiative transfer problems

Y Favennec[1], D Le Hardy[1], F Dubot[2], B Rousseau[1] and D R Rousse[2]
[1] LTN UMR CNRS 6607, Université de Nantes, France
[2] Chaire t3e, École de technologie supérieure, Montréal, Canada

E-mail: yann.favennec@univ-nantes.fr

Abstract. Inversion based on the radiative transfer equation (RTE) is generally highly CPU time consuming because the forward model itself is complicated to solve when the space dimension is greater than one, and because the inversion is based on a large number of forward model runs until convergence is reached. The goal of this paper is to set up some speed-up strategies specific of inversion when radiative transfer problems are dealt with. More specifically, the accurate identification of the volumetric radiative properties i.e. both the absorption and scattering coefficients is the objective of this study.

1. Statement of the Inverse Radiative Transfer Problem (IRTP)

The fine estimation of the volumetric radiative properties of semitransparent media remains a difficult task from the moment that their shapes and sizes do not match the features of the classical "1D homogeneous slab". From an experimental point of view, methodologies consist in collecting measurements by judisciously positionning sets of sources/detectors around the sample. This picture can be applied on BRDF/BTDF (bidirectional scattering distribution functions) characterization with an infrared spectrometer [1] as well as for optical tomography set-up (Figure 1). The inverse problem consits mathematically in minimizing the cost function which quantitatively measures the discrepancy between measurements $\breve{L}$ and related predictions L. For optical tomography, this cost function (1) mapping from $L_2(\mathcal{D}) \times L_2(\mathcal{D})$ to $\mathbb{R}$ uses, accordingl to the sketch of Figure 1, the specific euclidean norm.

$$j(\kappa, \sigma) \triangleq \frac{1}{2} \left\| L - \breve{L} \right\|_{2,\mathcal{X}}^2 \quad \text{with} \quad \|u\|_{2,\mathcal{X}} \triangleq \left(\sum_{k=1}^{K} \sum_{d=1}^{D} \sum_{r=1}^{R} u_k(\boldsymbol{x}_d, \boldsymbol{s}_r)^2 \right)^{\frac{1}{2}} \tag{1}$$

where the index k refers to the source (test) number, the index d refers to the detection node and the index r refers to the direction. For a given source configuration, say the kth one, the prediction $L_k(\boldsymbol{x}_d, \boldsymbol{s}_r)$ at the sensor d and direction r is given through the solution of the forward steady-state radiative transfer problem formulated as:

$$\left| \begin{array}{ll} \text{Find } L_k(\boldsymbol{x}, \boldsymbol{s}) \text{ such that } \forall k = 1, \cdots, K : & \\ (\boldsymbol{s} \cdot \nabla + \kappa + \sigma) L_k(\boldsymbol{x}, \boldsymbol{s}) = \sigma \oint_{\mathcal{S}^{n-1}} L_k(\boldsymbol{x}, \boldsymbol{s}) \Phi(\boldsymbol{s}, \boldsymbol{s}') \, \mathrm{d}\omega(\boldsymbol{s}') & \forall (\boldsymbol{x}, \boldsymbol{s}) \in \mathcal{D} \times \mathcal{S}^{n-1} \\ L_k(\boldsymbol{x}, \boldsymbol{s}) = L_k^0(\boldsymbol{x}) \times \mathbb{1}_{[\boldsymbol{x} \in \partial\mathcal{D}_k^0]} \times \mathbb{1}_{[\boldsymbol{s} = \boldsymbol{s}_k^0]} & \forall (\boldsymbol{x}, \boldsymbol{s}) \in \partial\mathcal{D} \times \mathcal{S}^{n-1} \end{array} \right. \tag{2}$$

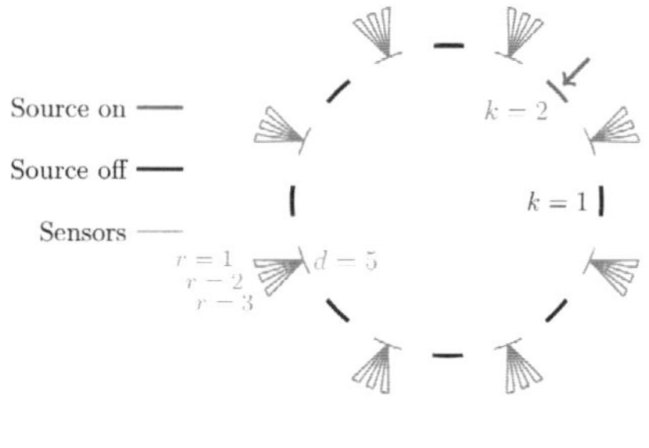

Figure 1. Sketch of the experiment. Only one source is on at once, while others are off. For each source configuration number $k = 1, \cdots, K$, the emerging intensity is measured on all sensors $d = 1, \cdots, D$ in different directions $r = 1, \cdots, R$.

2. Numerical recipes to speed-up IRTP

The inverse tomography problem which consists in minimizing (1) under the constraint that the forward model (2) is satisfied, necessitates heavy computation because (i) the forward model is an integro-differential equation which is known to be highly CPU-time consuming, (ii) each forward model is to be solved as many times as K to compute the set of predictions, (iii) the size of the control space is likely to be high, and the cost function is far from being quadratic, thus many optimization iterations are required even with efficient optimizers, and (iv) the inverse problem is ill-posed thus specific regularization is needed. In this study, the L-BFGS coupled with a quadratic line-search is used [2], the cost function gradient being computed through the adjoint-state formulation [3], the forward and backward RTEs being solved using the Discrete Ordinate Method (DOM) coupled with finite elements schemes based on specific formulations such as SUP-G or Discontinuous Galerkin, coupled with relaxation iterative algorithms [4]. The inverse radiative transfer problem (IRTP) still being very time consuming, additional tools to speed-up the global inversion have been developed.

The first family of tools for speeding-up the IRTP solutions is related to the solution of RTEs having in mind that inversion based on optimization is performed: for instance the forward RTE is solved differently is this context than if it was solved for a plain simulation. In the global inversion algorithm, several `for` loops are imbricated: the outer loop, say m, is the one related to the inversion optimizer, in between another loop is related to the source (test) configuration, say k, and the inner loops are related to the successive solutions of both the forward and backward RTEs, say p going from 1 to P until convergence is reached. The cost gradient and the direction of descent according to the L-BFGS are computed, and the line-search is performed. Between two successive optimization iterations, the properties variation $(\kappa, \sigma)^{(m)} - (\kappa, \sigma)^{(m-1)}$ is assumed to be moderate enough such that the initial guess at the optimization iteration m and for the source test k can be judiciously chosen equal to the solution given at the previous optimization iteration for the same source test:

$$L_k^{(m),(0)}(\boldsymbol{x}, \boldsymbol{s}) \leftarrow L_k^{(m-1),(P)}(\boldsymbol{x}, \boldsymbol{s}) \tag{3}$$

with a similar guess choice for the adjoint RTE. On the other side of the iterative algorithm, the stopping criterion used to stop the iterations while solving RTEs (that consists at the end in solving for each a linear system $\mathrm{Ax} = \mathrm{b}$) should also be chosen in accordance with the inverse settings. Usual stopping criteria are based on a norm of $\mathrm{x}^{(p)} - \mathrm{x}^{(p-1)}$ or better on a norm of the residual $\mathrm{r}^{(p)} = \mathrm{Ax}^{(p)} - \mathrm{b}$ using typically $\|\cdot\|_2$ or $\|\cdot\|_\infty$. Within the frame of inversion, it is useless to search tiny errors on residuals: the iterations should rather stop when errors on predictions become a fraction of the standard deviation $\sqrt{v}$ of measurement errors. With such a concept, the stopping criterion should be chosen as:

$$\left\| \mathrm{x}^{(p)} - \bar{\mathrm{x}} \right\|_{\infty, \mathcal{X}} < \epsilon_{\mathrm{x}} \triangleq \alpha \sqrt{v} \quad \text{with} \quad \|u\|_{\infty, \mathcal{X}} \triangleq \max_k \max_d \max_r |u_k(\boldsymbol{x}_d, \boldsymbol{s}_r)| \tag{4}$$

with v the variance of the measurement noise, and the coefficient α is chosen for instance to be equal to 10^{-1} or 10^{-2}. In practice $\bar{\mathrm{x}} = \mathrm{A}^{-1}\mathrm{b}$ is unknown, so the quantity $\left\|\mathrm{x}^{(p)} - \bar{\mathrm{x}}\right\|_{\infty, \mathcal{X}}$ is instead estimated by $\hat{e}^{(p)} = \delta_{p+1}^2 / (\delta_p - \delta_{p+1})$ with the notation $\delta_{p+1} \triangleq \|x^{(p+1)} - x^{(p)}\|_{\infty,\mathcal{X}}$. The RTE being linear with respect to the Dirichlet condition, the stopping criterion for the adjoint problem is chosen to be the number of iterations of the forward model at convergence.

The second family of tools for speeding-up the IRTP consists in considering a different scaling than those found in the literature. Indeed, a scaling is necessary due to the fact that two very distinct physical properties $\kappa(\boldsymbol{x})$, $\sigma(\boldsymbol{x})$ are to be recovered (both act differently on measures and both magnitudes are different). Often, the scaling is performed on the cost function gradient following [5] where it was suggested to use the information on gradients computed before the first iteration. Such gradient scaling lead to consider, with $c_\alpha = 0.05 \times \|\alpha^{(0)}\|_{L_\infty(\mathcal{D})} / \|\nabla_\alpha j^{(0)}\|_{L_\infty(\mathcal{D})}$:

$$\tilde{\nabla} j^{(m)} = c_\kappa \nabla_\kappa j^{(m)} + c_\sigma \nabla_\sigma j^{(m)} \tag{5}$$

This approach has been used for instance by [3] and many others. Elsewhere, in [6], another approach using adimensionalization using priors was proposed so that the estimation finally was performed on $\tilde{\kappa} = \kappa/\kappa_{\text{ap}}$ and $\tilde{\sigma} = \sigma/\sigma_{\text{ap}}$. It was shown that this adimensionalization approach gave similar convergence results as the gradient scaling method [5] in a DOT problem. It is here suggested to verify that this adimensionalization approach is indeed the optimal approach. To do so, we combine the adimensionalization approach with a gradient scaling but with coefficient c_κ, c_σ determined again before the first optimization iteration but this time with (ϵ being small enough): $\gamma = \left(\sum_{\alpha=\kappa,\sigma} 0.05 \times \|\alpha^{(0)}\|_{L_\infty(\mathcal{D})} / \|\nabla_\alpha j^{(0)}\|_{L_\infty(\mathcal{D})}\right)^{1/2}$, $c_\kappa = \gamma \cos\theta^*$, $c_\sigma = \gamma \sin\theta^*$, $\theta^* = \arg\min_{\theta \in]0,\frac{\pi}{2}[} j(\gamma^0 - \epsilon \nabla j^0)$, and verify numerically that the obtained optimal θ^* corresponds to the adimensionnalization.

The third family of tools for speeding-up the IRTP consists in using regularization because the IRTP is ill-posed. Among numerous regularization strategies, one that counteracts the effect of the measurement noise where its effect is the greatest (ie close to the sensors) is developped. This is an extension of the Sobolev gradient [7] in which the Sobolev parameter decreases in an exponential way with the distance to the sensors. The Sobolev space $H^{1(\ell(\boldsymbol{x}))}(\mathcal{D})$ is defined based on the inner product definition $(u,v)_{H^{1(\ell(\boldsymbol{x}))}} = (u,v)_{L_2(\mathcal{D})} + \ell^2(\boldsymbol{x}) (\nabla u, \nabla v)_{L_2(\mathcal{D})}$ with $\ell^2(\boldsymbol{x}) = \exp\left(-\gamma d^2(\boldsymbol{x})\right)$, $d(\boldsymbol{x})$ being the distance betweeen $\boldsymbol{x}$ and its closest sensor and γ is the new regularization parameter. With such an inner product definition, the ordinary cost function gradient $\nabla_\alpha j$ being computed for both $\alpha = \kappa$, σ, the Sobolev cost function gradient $\nabla_\alpha^{H^{1(\ell(\boldsymbol{x}))}} j$ is given by the solution of, for $\alpha = \kappa$, σ:

$$\left| \begin{array}{ll} \left(1 - \ell^2(\boldsymbol{x})\Delta\right) \nabla_\alpha^{H^{1(\ell(\boldsymbol{x}))}} j = \nabla_\alpha j & \boldsymbol{x} \in \mathcal{D} \\ \nabla\nabla_\alpha^{H^{1(\ell(\boldsymbol{x}))}} j \cdot \boldsymbol{n} = 0 & \boldsymbol{x}\partial\mathcal{D} \end{array} \right. \tag{6}$$

The combination of the three presented tools (guess initialization and appropriate convergence criteria within Gauss–Seidel iterative schemes, new gradient scaling, and the use of a new space-dependent Sobolev inner product when extracting the cost gradients) are the new ingredients developped to tremendously save-up computational time for solving inverse radiative transfer problems. Numerical results will be presented at the conference.

References

[1] Randrianalisoa J H, Baillis D and Pilon L 2006 *Journal of thermophysics and heat transfer* **20** 871–883
[2] Liu D and Nocedal J 1989 *Mathematical Programming: Series A and B* **45** 503–528
[3] Balima O, Favennec Y and Rousse D 2013 *Journal of Computational Physics* **251** 461–479
[4] Le Hardy D, Favennec Y and Rousseau B 2015 Iterative and fem methods to solve the 2d radiative transfer equation with specular reflexion *CTRPM congress, Albi*
[5] Klose A 2001 *Optical tomography based on the Equation of Radiative Transfer* Ph.D. thesis Department of Physiscs Freie Universität, Berlin Germany
[6] Dubot F, Favennec Y, Rousseau B and Rousse D (submitted) *IJTS*
[7] Protas B 2008 *Journal of Computational Physics* **227** 6490–6510

Journal of Physics: Conference Series **676** (2016) 012007 doi:10.1088/1742-6596/676/1/012007

Radiative, conductive and convective heat-transfers in a single Monte Carlo algorithm

Richard Fournier[a,b], Stéphane Blanco[a,b], Vincent Eymet[c], Mouna El Hafi[d], Christophe Spiesser[d]

[a] Université de Toulouse ; UPS, INPT ; LAPLACE (Laboratoire Plasma et Conversion dEnergie) ; 118 route de Narbonne, F-31062 Toulouse cedex 9, France.
[b] CNRS ; LAPLACE ; F-31062 Toulouse, France.
[c] Méso-Star SAS, 8 rue des pêchers, 31410 Longages, France
[d] Université de Toulouse, Mines-Albi, UMR-CNRS 5302, Centre RAPSODEE, Campus Jarlard, F-81013 Albi CT cedex 09, France

E-mail: richard.fournier@laplace.univ-tlse.fr

Abstract. It was recently shown that null-collision algorithms could lead to grid-free radiative-transfer Monte Carlo algorithms that immediately benefit of computer-graphics tools for an efficient handling of complex geometries [1, 2]. We here explore the idea of extending the approach to heat transfer problems combining radiation, conduction and convection. This is possible as soon as the model can be given the form of a second-kind Fredholm equation. In the following pages, we show that this is quite straightforward at the stationnary limit in the linear case. The oral presentation will provide corresponding simulation examples. Perspectives will then be drawn concerning the extension to non-stationnary cases and non-linear coupling.

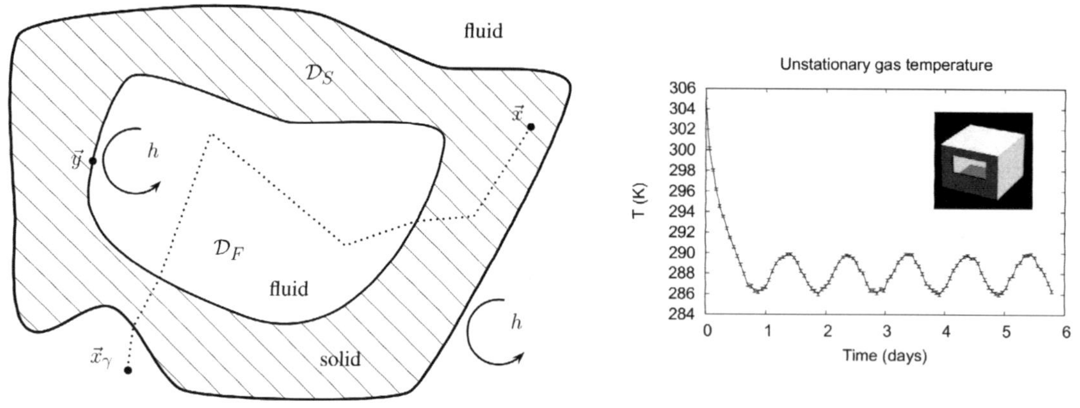

Figure 1. Fluid and solid domains (left). Academic example of a Monte Carlo simulation dealing with radiative, conductive and convective heat-transfers, for opaque solids and transparent fluids (internal air-tempertaure for a sinusoidal external forcing, right).

Journal of Physics: Conference Series **676** (2016) 012007 doi:10.1088/1742-6596/676/1/012007

We consider a system constituted of solid and fluid connex domains, noted $\mathcal{D}_S$ and $\mathcal{D}_F$ respectively, of boundary $\partial\mathcal{D}_S$ and $\partial\mathcal{D}_F$. The heat transfer modes are conduction/radiation in the solid, and convection/radiation in the fluid. Both the solid and the fluid are semi-transparent and grey. Density ρ, specific heat capacity C, conductivity λ and absorption coefficient k_a in the solid are all heterogeneous but independant of temperature. The fluid is assumed perfectly mixed in each connex domain : ρ, C and k_a are therefore homogeneous. They are also independent of temperature. Convective heat transfer is resumed to an heat exchange coefficient h at each location at the solid boundary, and an incoming/outcoming flow of volumetric flow-rate ϕ, the incoming flow being of a known temperature θ_N. These three quantities are independent of the solid and fluid temperatures. Radiation is linearised around a reference temperature θ_{ref}, defining a volumetric radiative heat transfer coefficient $\zeta = 16k_a\sigma\theta_{ref}^3$, where σ is the Stefan-Boltzmann constant. Without precising the initial conditions, the resulting heat-transfer model is linear in temperature:

$$\left.\begin{array}{rr} \vec{x} \in \mathcal{D}_S \ : \ \rho C\dfrac{\partial\theta}{\partial t} = & -\vec{\nabla}.\left(-\lambda\vec{\nabla}\theta\right) + \zeta(\theta_R - \theta) \\ \vec{x} \equiv \vec{y} \in \partial\mathcal{D}_S \ : & -\lambda\vec{\nabla}\theta.\vec{n} = h(\theta_F - \theta) \end{array}\right\} \quad \text{solid} \qquad (1)$$

$$\left.\begin{array}{rl} \rho C\mathcal{V}_F\dfrac{\partial\theta}{\partial t} = \rho C\phi(\theta_N - \theta) & + \ \zeta\displaystyle\int_{\mathcal{D}_F}(\theta_R(\vec{x}_R,t) - \theta)d\vec{x}_R \\ & + \displaystyle\int_{\partial\mathcal{D}_S} h(\vec{y},t)(\theta_S(\vec{y},t) - \theta)d\vec{y} \end{array}\right\} \quad \text{fluid} \qquad (2)$$

$$\left.\theta_R = \int_{\mathcal{D}_\Gamma} p_\Gamma(\gamma)d\gamma\ \theta(\vec{x}_\gamma, t - t_\gamma)\right\} \quad \text{solid or fluid} \qquad (3)$$

where θ is temperature, t is time, $\vec{x}$, $\vec{x}_R$ (and later $\vec{x}_G$) are locations within the solid or fluid domains, and $\vec{y}$ (and later $\vec{y}_G$) is a location at the boundary. At a boundary location, θ_S and θ_F are respectively the solid and fluid temperatures; $\vec{n}$ is the unit normal heading toward the solid. The radiative temperature θ_R is defined using the space $\mathcal{D}_\Gamma$ of all trajectories γ of photons emitted at $\vec{x}$, until their absorption at $\vec{x}_\gamma$. The probability density $p_\Gamma(\gamma)$ of trajectory γ reflects the statistical pictures associated to the radiative transfer equation and θ_R is therefore simply the average value of the temperature at absorption locations, for photons emitted at $\vec{x}$.

As the problem is linear, its general solution can be expressed using an integral formulation involving unstationnary Green functions. We here only give the stationnary solution:

$$\left.\begin{array}{rl} \vec{x} \in \mathcal{D}_S \ : \ \theta(\vec{x}) = & \displaystyle\int_{\mathcal{D}_S} d\vec{x}_G\ G_{S,stat}(\vec{x};\vec{x}_G)\ \theta_R(\vec{x}_G) \\ & + \displaystyle\int_{\partial\mathcal{D}_S} d\vec{y}_G\ G_{S,stat}(\vec{x};\vec{y}_G)\frac{h(\vec{y}_G)}{\zeta(\vec{y}_G)}\ \theta_F(\vec{y}_G) \end{array}\right\} \quad \text{solid} \qquad (4)$$

$$\left.\theta = \frac{\rho C\phi\theta_N \ + \ \zeta\int_{\mathcal{D}_F}\theta_R(\vec{x}_R)d\vec{x}_R \ + \ \int_{\partial\mathcal{D}_S} h(\vec{y})\theta_S(\vec{y})d\vec{y}}{\rho C\phi \ + \ \zeta\mathcal{V}_F \ + \ \int_{\partial\mathcal{D}_S} h(\vec{y})d\vec{y}}\right\} \quad \text{fluid} \qquad (5)$$

$$\left.\theta_R = \int_{\mathcal{D}_\Gamma} p_\Gamma(\gamma)d\gamma\ \theta(\vec{x}_\gamma)\right\} \quad \text{solid or fluid} \qquad (6)$$

where $\mathcal{V}_F$ is the volume of $\mathcal{D}_F$, and $G_{S,stat}(\vec{x};\vec{x}_G)$ is the value of the stationnary Green function, at $\vec{x}$ within the solid, for a source at $\vec{x}_G$ within the same solid domain.

These integrals can each be interpreted as expectations of random variables and their combination allows to interprete the temperature at any location, in the solid or in the fluid, as the expectation of a random function $\Theta(\vec{x})$. The Monte Carlo algorithm then evaluates $\theta(\vec{x})$ as the average of a large number of Θ samples [3]. The definition of Θ combines six random functions of location: A_N, A_R and A_NC for the selection a source type (incoming flow, radiation or convection), $\vec{X}_G$ for the sampling of a location within the solid or fluid domain, $\vec{Y}_G$ for sampling of a location at the boundary and Θ_R that samples of an optical path and returns the temperature at its extremity.

$$\left.\begin{aligned} \theta(\vec{x}) &= \mathscr{E}\left[\Theta(\vec{x})\right] \\ \Theta(\vec{x}) &= A_N(\vec{x})\theta_N + A_R(\vec{x})\Theta_R(\vec{X}_G(\vec{x})) + A_C(\vec{x})\theta_C(\vec{Y}_G(\vec{x})) \end{aligned}\right\} \quad \text{solid or fluid} \tag{7}$$

As $\Theta_R(\vec{X}_G(\vec{x}))$ and $\theta_C(\vec{Y}_G(\vec{x}))$ involve the solid and fluid temperatures at random locations, these temperatures can again be replaced by the expectation of Θ at the corresponding locations and the Monte Carlo algorithm becomes recursive, just as for simulation of multiply scattered radiation. The algorithm stops when the required temperature is known, either because this corresponds to the incoming flow and θ_N is known, or because the sampled location is at the boundary. This recursivity is a typical consequence of the combination of equations 4, 5 and 6 leading to a second-kind Fredholm equation.

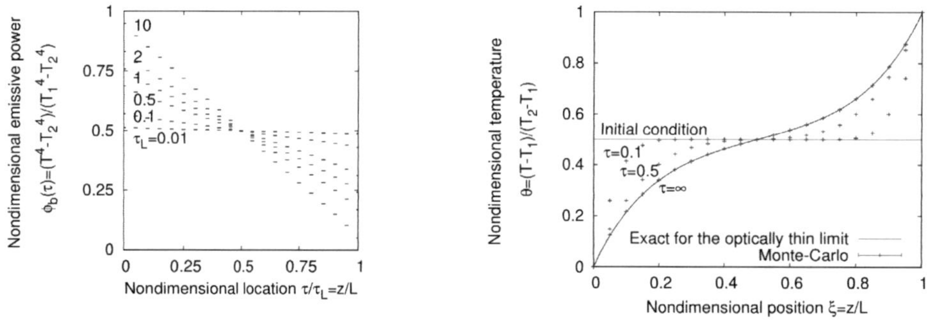

Figure 2. Academic 1D examples involving semi-transparent solids: radiative "equilibrium" (left); radiation coupled with non-stationnary conduction (right). Notations are from [4] page 460 and page ??.

[1] Galtier M, Blanco S, Caliot C, Coustet C, Dauchet J, El Hafi M, Eymet V, Fournier R, Gautrais J, Khuong A, Piaud B and Terree G 2013 *J. Quant. Spectr. Rad. Transfer* **125** 57–68 ISSN 0022-4073

[2] Eymet V, Poitou D, Galtier M, El Hafi M, Terree G and Fournier R 2013 *J. Quant. Spectr. Rad. Transfer* **129** 145–157 ISSN 0022-4073

[3] Delatorre J, Baud G, Bezian J J, Blanco S, Caliot C, Cornet J F, Coustet C, Dauchet J, El Hafi M, Eymet V, Fournier R, Gautrais J, Gourmel O, Joseph D, Meilhac N, Pajot A, Paulin M, Perez P, Piaud B, Roger M, Rolland J, Veynandt F and Weitz S 2014 *Solar Energy* **103** 653–681 ISSN 0038-092X

[4] Modest M F 1993 *Radiative Heat Transfer* (McGraw-Hill, Inc.)

Journal of Physics: Conference Series **676** (2016) 012008 doi:10.1088/1742-6596/676/1/012008

Lab-scale study of radiative fluxes received from a fire front

J Gérardin, A Marchand, N Trévisan, A Collin, Z Acem, P Boulet [1]
LEMTA, CNRS, Université de Lorraine
2 av. de la Forêt de Haye, TSA 60604 – 54518 Vandœuvre-lès-Nancy Cedex, France
E-mail: pascal.boulet@univ-lorraine.fr

Abstract. Numerical simulations of radiation received from a fire front were carried out in a situation of laboratory-scale fire. The fire front was determined at different instants based on camera images taken during a real experiment, and predicted in using a "small world network" propagation model. The fluxes were computed using either a ray tracing method with EDStaR, or a home-made reciprocal Monte Carlo method. Results were compared with available flux measurements using radiative heat flux gauges.

1. Introduction

Radiative transfer is one of the leading mechanisms in fire propagation. It is also a key transfer mode when evaluating the possible hazards due to fire spread. The building rules and the size of the area free of vegetation must be determined considering the impact of radiative fluxes on the houses. This is of particular interest for countries where urban areas are inter-connected with forest areas, like in France. As a preliminary step before real scale evaluations, experiments were carried out at laboratory scale on burning tables. This was indeed a first calibration step. Fire spread was studied on a configuration in which flux sensors were placed on the faces of a cube centred in a circular vegetation-free area. The fire was seen to reach and spread around the cube featuring a house and the fluxes were registered. The goal of the numerical work conducted on the same configuration was to simulate the radiative transfer and to compare the predicted fluxes on the different faces of the cube with the experimental data. For this purpose, the fire front was extracted as a function of time from camera recordings carried out during the experiment. Then a ray tracing simulation was performed using EDStaR, simulating the radiation from this fire front. In parallel a complete simulation of fire propagation was conducted using a small world network model and comparisons were carried out on the fluxes received by the target. The present communication briefly presents the setup, the simulations and first comparisons between experimental data and simulation results.

2. An overview of the experimental setup PROMETHEI

Experiments were carried out at the experimental platform PROMETHEI, which was designed by the LEMTA in Nancy to perform fire propagation tests at the laboratory scale. Burning tables 3.8 m x 3 m in size were used. The fuel was excelsior, with a vegetation load of 500 g/m^2. The fire was ignited on a line. A fuelbreak was considered as a circular surface with diameter 1 m and its center is located at 1.9 in X axis and 1.7 m in Y axis. A cubic structure was located at the center of this area, supporting

[1] To whom any correspondence should be addressed.

Journal of Physics: Conference Series **676** (2016) 012008 doi:10.1088/1742-6596/676/1/012008

radiative heat flux gauges on each face (see Figure 1). Visible cameras (one on the rear side and one on the front side) were used to identify the fire front position as a function of time. Details on the setup and on the experimental facility can be found in [1]. An example of fire front reconstruction by SWN model is presented in Figure 2 (experimental data with black and blue dots compared to fire propagation model predictions with red symbols) when the fire is propagating around the fuelbreak area.

Figure 1: Picture of the setup after fire ignition

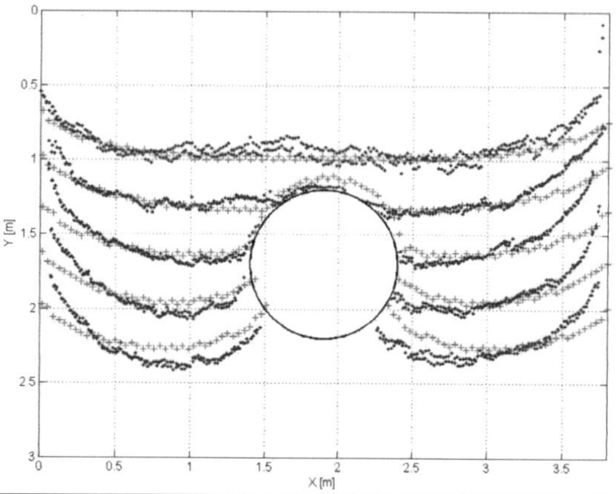

Figure 2: Example of a fire front shape reconstruction and identification as a function of time.

3. Numerical simulations

Based on the fire front reconstruction, simulations were performed for the radiative transfer with EDStaR (*Environnement de Développement Statistique Radiative*) taken from PBRT, which is a physically based rendering system based on the ray-tracing algorithm [2]. It is originally used to simulate the effect of light on a described geometry. A group of researchers from laboratories RAPSODEE, LAPLACE and PROMES, used the environment of PBRT to create EDStaR [3], with the aim of performing radiative transfer calculations. This numerical toolbox also calculates statistical errors and it is optimized for parallel calculations (MPI). The geometry of the experimental setup was computed inside EDStaR, to represent the vegetation, the box with the radiative flux detectors and a flame which follows the shape of the front fire. The flame was firstly considered as a radiant panel. A view factor was then calculated between the flame and each radiative flux detectors, with the use of the ray-tracing algorithm of EDStaR.

In parallel, a complete prediction of the fire propagation was done with a small-world network code [1] combined with a reciprocal Monte Carlo method simulating the radiation received by the heat flux gauges from the volume of flame. Input data required for the simulation were taken from previous experiments by the authors [1,4], namely flame temperature 1400 K, flame absorption coefficient 0.2 m^{-1} and flame height 50 cm. A sensitivity analysis is of course required in order to check the exact influence of these input data (work in progress).

4. Results and discussion

The fire propagation experiments have been repeated two times (Run #1 and Run #2) order to analyze the reproducibility of experimental results. The experimental radiative fluxes obtained are plotted in

Figure 3. The relative discrepancies can reach 50% (for the Front side detector). An average of these data between Run #1 and Run #2 is here used for the comparison with the numerical results.

Radiative fluxes were computed for the different detectors (front, lateral and top detectors). They are plotted in Figure 4 for the flame height of 50 cm, which was found to yield the lowest discrepancy between averaged experimental data and numerical results.

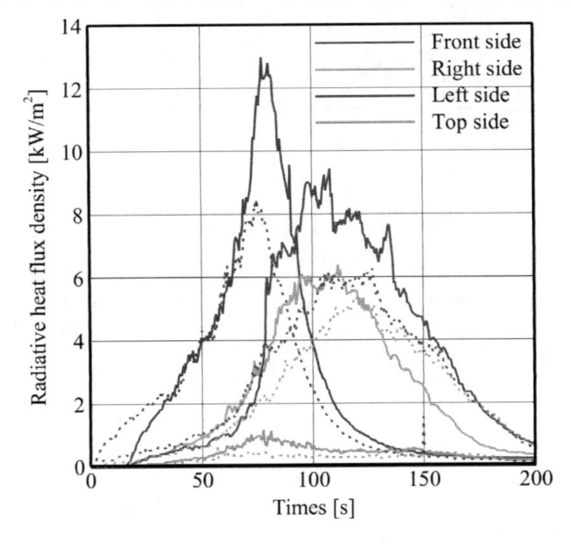

Figure 3: Radiative flux on each detector (solid lines – Run #1; dotted lines – Run #2) with a fire front of 50 cm in height.

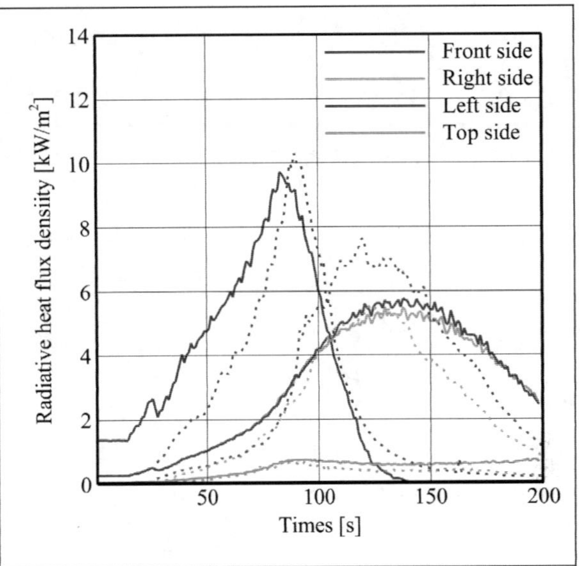

Figure 4: Comparison between radiative fluxes (solid lines - Simulation results; dotted lines – Average of the experimental results).

These are only preliminary results but it seems that the main trends are captured as the flame propagates. This method of flux simulation will be further extended in order to allow a comparison for all the fire propagation experiment.

5. Conclusions

Simulations were carried out on the fluxes received from a fire front combining a fine reconstruction of the fire front location with a ray tracing method for the radiative flux evaluation. Preliminary results show the possibility to obtain the radiative fluxes with sensitivity data for the flame geometry in particular. Further tests will be conducted for optically thin or tick flames, instead of a solid flame, involving a better description of the true flame geometry and of its emission.

References

[1] A. Marchand, N. Trévisan, A. Collin and P. Boulet. Experimental and numerical study of fire behaviour: Effects of the width on the rate of spread. In proceedings of VIIth International Conference on Forest Fire Research - Coimbra, Portugal, November 17-20th, 2014.
[2] M. Pharr and G. Humphreys. Physically based rendering: from theory to implementation, Elsevier, 2004.
[3] ED STAR. Retrieved December 14, 2014, from http://edstar.lmd.jussieu.fr/
[4] G. Parent, Z. Acem, S. Lechêne, P. Boulet. Measurement of infrared radiation emitted by the flame of a vegetation fire International Journal of Thermal Sciences, 49, pp. 555-562, 2010.

Towards the development of simple methods for determining normal absorptances of open-cell foams based on opaque materials

Simon Guévelou, Benoit Rousseau, Gilberto Domingues
CNRS, LTN UMR 6607, Ecole Polytechnique de l'Université de Nantes, Rue Christian Pauc, 44306 Nantes Cedex 3, France.

Jérôme Vicente
Aix-Marseille Université, CNRS, IUSTI UMR 7343, 13453 Marseille Cedex 13, France
E-mail: `simon.guevelou@univ-nantes.fr`

Abstract. The knowledge of the normal spectral absorptances of open-cell foams used as volumetric solar receivers is required to finely compute their thermal efficiencies. For this purpose, absorptances of a set of virtual open-cell foams of varying porosities, beforehand generated by using a numerical generator, are computed thanks to a ray tracing code. This work details the contribution of the intrinsic optical properties of the solid phase to the normal spectral absorptance of open-cell foams, through the statistical analysis of both the path and the events undergone by a ray, which is permitted by the ray tracing code. This study allows us to propose a robust and linear relationship that links the normal spectral absorptance to the porosity and the intrinsic optical properties of the solid phase, which is considered as optically thick.

1. Introduction

The production of electric power from solar thermal energy, specifically technologies based on concentrated solar power, is nowadays growing at a fast pace. The optimization of heat conversion that occurs within the receiver, on which solar radiation is concentrated, is crucial in order to obtain higher efficiencies. The use of volumetric porous absorbers, based on open-cell silicon carbide foams (SiC), permits to heat the outside air that is pumped directly through the porous media at high temperatures (1000 - 1400 K). In particular, the accurate determination of the radiative exchanges within SiC open-cell foams requires the finding of the links between their textural features and their radiative properties. Several computational approaches, allowing the characterization of the both the homogenized absorption and scattering coefficients as the scattering phase function, were developed to fulfill this objective and were recently summarized by Baillis et *al.* [1]. The present paper focuses on the prediction of the normal spectral absorptance, A, which is especially needed to predict the temperature profiles that occurs within insolated foams [2]. Contrarily to the previous homogenized radiative properties, the links between foam's textural parameters and its absorptance are less known [3]. Thus, a ray tracing code and a numerical foam generator [4, 5], have been combined.

By using these numerical tools, a set of foams with varying porosity was generated to study the impact of porosity on the absorptance. Akolkar and Petrasch [3], made a study with fixed solid interface reflectivity values and for volumes smaller than representative elementary volume (REV) for A. The current work goes further by attempting to establish a robust and simple linear relationship that directly links A to the porosity of the foam, p, and to $\tilde{n}$ (*i.e.* the complex index of refraction $\tilde{n} = n + ik$, with n and k being respectively the refractive and absorptive index), within volumes higher than REV.

2. Set of foams with varying porosity

The foam generator allows to generate open-cell foams whose textural features (porosity, specific surface, pore size distribution mean cell connectivity) are close to those measured on the real SiC foam of interest. The measurement of those textural features was realized thanks to the free software iMorph. A set of four cubic foam samples, whose PPI (number of Pore Per Inch) value are equal to 23 and porosities values are equal to 0.3, 0.5, 0.7 and 0.9, was generated.

3. Ray tracing procedure and ray transport statistical analysis

The ray tracing procedure allows the computation of the normal hemispherical reflectance, R_{NH}, and transmittance, T_{NH}. The comparizon between the thermal radiation wavelength and the pores and struts sizes allows the use of geometrical optics approximation. A large number of rays $N_{rays} = 10^6$, whose directions are collimated to the normal of the foam sample, are launched within a circular spot. Each ray is independently tracked through the 3D reconstruction as follows: void phase is considered as non-participating media and, at each local interaction with the solid phase, a local reflectivity is computed via a specular model according to Fresnel's law. This model needs the knowledge of $\tilde{n}$ of the solid phase constituting the struts. In this work, we will use a set of couple (n, k) which can guarantee a modulation of the local reflectivity. With these couples, we can cover the optical properties of SiC and other opaque material for the considered thicknesses of the struts. Once R_{NH} and T_{NH} are computed, normal spectral absorptance A can be obtained since $A = 1 - R_{NH} - T_{NH}$. A thorough statistical analysis of ray transport, allowing the determination of the total number of ray impacts on the solid phase, and the related mean local reflectivity values, is made for each ray path.

4. Results

A computations were realized on the same cubic foam numerical sample of varying porosities, with a wide range of arbitrary n and k values ($n \in$ [1.5 - 2.5] and $k \in$ [0.5 - 8]). The volumes of the samples were larger than the REV determined for A, implying that $T_{NH} = 0$. For each computation, the mean local reflectivity ρ_m was computed. ρ_m is defined here by the average, over each ray path within an entire ray tracing computation, of the successive local reflectivity values that are given by the Fresnel's law. Along a single ray path on which it undergo N_{imp} impacts, the mean local reflectivity related to this ray path, $\rho_{m,path}$, is given by:

$$\rho_{m,path} = \frac{1}{N_{imp}} \sum_{1}^{N_{imp}} \rho_i \tag{1}$$

with ρ_i being the computed reflectivity at the ith impact. ρ_m does not take into account energy loss due to the multiple reflections of a ray and must be considered here as a statistical mean local reflectivity related to the distribution of the orientations of the facets encountered along the paths of a ray tracing computation.

Fig. 1 shows A as function of p, n and k for several arbitrary n and k values. A grows linearly with porosity, and each slope (denoted here S) related to a specific n and k couple, and so on, to a particular ρ_m value, were computed. For n and k values that give high values of ρ_m, the lower

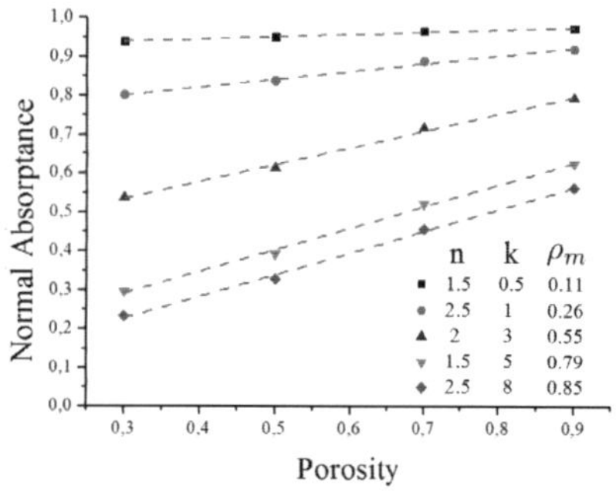

Figure 1. Normal absorptance as function of the porosity for abitrary n and k values.

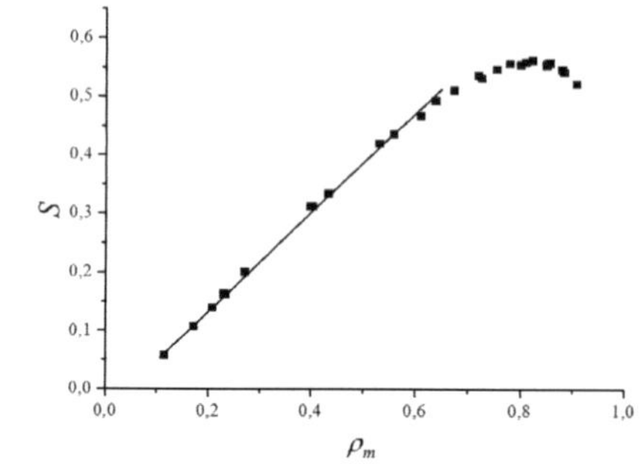

Figure 2. Slopes S (of A as function of p) as function of the mean local reflectivity ρ_m.

is A, and the stronger is the influence of p. The values of the slopes S are plotted as function of the ρ_m corresponding to the same A computations on Fig. 2. It is shown that S depends linearly of ρ_m values on a wide range of ρ_m values. As ρ_m is a function of n, k and the incident angle distribution, this feature allows to define S values also as a function of the intrinsic optical properties of the solid phase constituting the foam. For highly reflective materials ($\rho_m > 0.6$) the value of S decreases. This results permits to show that (for $\rho_m < 0.6$),

$$A = \frac{3}{4} \times \rho_m \times p + A_{strut}, \tag{2}$$

where ρ_m depends on n, k and on the angle distribution of the facets, and with A_{strut}, the absorptance of a strut. Special attentions must be paid now to take into account the influence of the multiple reflections when computing ρ_m values.

5. Conclusion

A linear relationship that links A to porosity and intrinsic optical properties of the solid phase constituting the struts was found in this work, for volumes higher than REV. It is shown that in this case, A can be simply deduced from measurement performed on a dense slab with similar chemical composition. This result relies on a thorough statistical analysis of the transport of rays which permits us to go further in the prediction of radiative properties of foams. However, the influence of the orientation distribution of the facets constituting the 3D reconstruction on this results must be quantified.

References

[1] D. Baillis, R. Cocquard, J. Randrianalisoa, L. Dombrovsky, and R. Viskanta, "Thermal radiation properties of highly porous cellular foams," *Special Topics and Reviews in Porous Media*, vol. 4, no. 2, pp. 111–136, 2013.

[2] A. Kribus, Y. Gray, M. Grijnevich, G. Mittelman, S. Mey-Cloutier, and C. Caliot, "The promise and challenge of solar volumetric absorbers," *Solar Energy*, vol. 110, pp. 463–481, 2014.

[3] A. Akolkar and J. Petrasch, "Tomography based pore-level optimization of radiative transfer in porous media," *International Journal of Heat and Mass Transfer*, vol. 54, pp. 4775–4783, 2011.

[4] B. Rousseau, J.-Y. Rolland, P. Echegut, E. Brun, and J. Vicente, "Numerical prediction of the radiative behavior of metallic foams from the microscopic to macroscopic scale," in *Journal of Physics : Conference Series, Eurotherm Conference No. 95, Nancy, France*, vol. 369, IOP, 2012.

[5] S. Guévelou, B. Rousseau, G. Domingues, J. Vicente, C. Caliot, and G. Flamant, "Identification of the radiative properties of α-SiC foams realistically designed with a numerical generator," in *Proceedings of the 15th International Heat Transfer Conference, Kyoto, Japan*, 2014.

Journal of Physics: Conference Series **676** (2016) 012010 doi:10.1088/1742-6596/676/1/012010

Radiative characterization of ceramic foams with microporosity

S Haussener[1], J. Randrianalisoa[2]
[1] Ecole Polytechnique Fédérale de Lausanne, Institute of Mechanical Engineering, Lausanne, Switzerland
[2] Groupe de Recherche en Sciences Pour l'Ingénieur (GRESPI), EA 4694, Université de Reims Champagne-Ardenne, Campus du Moulin de la Housse, F-51687, Reims, France

E-mail: sophia.haussener@epfl.ch, jaona.randrianalisoa@univ-reims.fr

Abstract. The porous morphology of ceramic foams can significantly influence its heat and mass transport phenomena. Ceramic foams with dual-scale porosity provide flexibility for tailoring the coupled transport characteristics for enhanced performance. We numerically characterized the radiative transport in porous ceria foams with dual-scale porosity, i.e. exhibiting pores in the millimeter range in the micrometer range. Ceria can act as a catalyst-equivalent in high temperature thermochemical reactions for the direct synthesis of solar fuels and its bulk material properties vary significantly with wavelength. The methodology used is based on Monte Carlo methods for the solution of the volume-averaged radiative transfer equations for the determination of macroscopic optical properties such as reflectance or transmittance of a 1D slab. The exact millimeter-scale structure was incorporated by effective transport properties obtained through collision-based Monte Carlo methods. The micrometer-range strut porosity was incorporated using Mie theory and assuming independent scattering. The results allow for guiding the synthesis of ceramic foams with dual-scale porosity for enhanced radiative transport characteristics.

1. Introduction

Ceramic foams play an important role in high temperature applications as radiative burners or absorber and reactant for solar thermochemical fuel processing. The latter provide a direct pathway for solar fuel processing using ceramics such as ceria as a catalyst for thermochemical water splitting [2]. The ceramic foam acts simultaneously as direct solar absorber, heat exchanger, reactant and reaction site. The accurate characterization of the radiative transfer in these complex structures (shown in figure 1.a) is required for an in-depth understanding of the coupled heat and mass transfer, and chemical reaction. Tomography-based approaches have been used to accurately determine the heat and mass transport in complex structures at scales above 1μm [5], and to tailor the foam morphology for increased performance [10]. A further performance improvement by increasing the specific surface through the introduction of dual-scale porosity, i.e. small scale pores within the struts of the porous foam [4] (shown in figure 1.b), has been experimentally demonstrated. The performance increase results from enhanced radiation absorption and increased specific surface for the chemical reaction. We aim at providing a methodology which allows for the characterization of ceramic foams exhibiting dual-scale porosity, i.e. macroscopic pores in the range of mm, and microscopic pores in the range of μm.

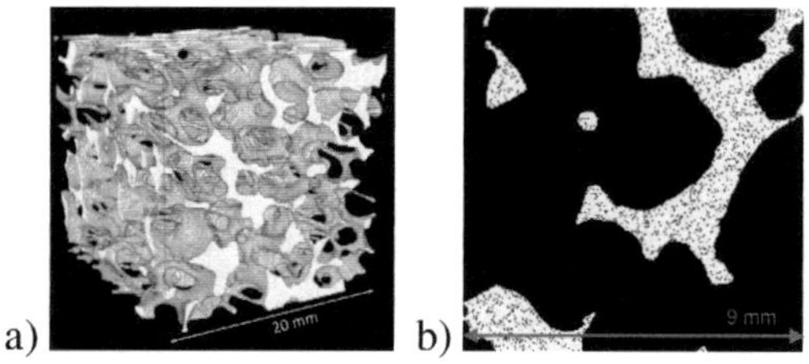

Figure 1. a) 3D rendering of a ceria ceramic foam with porosity of 82%, obtained by X-ray tomography with resolution of 35.7 μm [10], and b) cut through a quarter of the sample indicating the 10% strut-porosity with micro-pore radius of 18 μm.

2. Problem statement

The media considered were statistically isotropic and homogeneous porous media obtained by the replica method [9], and composed of an isotropic, percolating, opaque or semitransparent solid phase, and a radiatively non-participating, stagnant fluid phase. The radiative bulk properties of ceria, i.e. its spectral complex refractive index, are given in literature [8]. The investigations were based on the digitalized morphological data of a sample with porosity of 82%. The porosity of this sample was then artificially decreased by increasing the strut diameter by dilation operations [10]. We considered 2 samples of 82% and 60% porosities. The microscopic pores in the struts were artificially added to the foam struts and spanned pore sizes of 10 to 40 μm at a volume fraction in the solid phase of 0 to 40%. Geometrical optics was assumed for the calculation of the macroscopic properties as the smallest dimensions of the solid structure (about 0.4 mm) was significantly larger than the wavelength range considered. Diffraction was neglected as for large particles it is predominantly in forward direction within a small solid angle. The radiative characteristics of struts containing microscopic pores was obtained by including the contribution of ceria absorption and the scattering by pores in the framework of independent scattering theory [1]. Possible dependent scattering effects for highly porous struts were neglected. Scattering efficiency factors and phase functions by small pores were calculated using generalized Mie theory [11], accounting for the possibly absorbing ceria surrounding the pores.

3. Methodology

The methodology is based on volume-averaging approaches used for the characterization of the radiative transfer in multi-phase media. Two coupled radiative transfer equations (RTEs) are solved [7, 6, 3] for an infinite 1D slab. The boundary conditions for the slab are reflecting at the lateral walls, absorbing environment at the outlet, and collimated irradiation at the inlet. A path-length based Monte Carlo method is used for the solution of the volume-averaged RTEs. The macroscopic radiative properties, reflectance and transmittance of the 1D slab, are calculated based on the volume-averaged intensity vector field in the two phases and its integration over all solid angles of the positive and negative half-sphere at the inlet and outlet. The effective properties required for the volume-averaged RTEs, namely absorption coefficients for the fluid and solid phases, four scattering coefficient accounting for refraction and reflection at both sides of the interface boundaries, and extinction coefficients for the fluid and solid phase, and four scattering phase functions accounting for the angular distribution of the refraction and reflection at both sides of the interface boundaries, are determined using a collision-based Monte Carlo method in the 3D numerical sample [7, 6, 3]. The bulk properties accounting for the small-scale porosity within the struts are calculated using the BHMIE algorithm [1].

4. Results

The effective radiative properties for the void and solid phases (subscripts 1 and 2, respectively), namely extinction coefficients (β_1, β_2), scattering albedos (ω_{11}, ω_{12}, ω_{21}, ω_{22}), and scattering function asymmetry factors (g_{21}, g_{22}), are depicted in figure 2. The macroscopic optical properties, namely transmittance and reflectance of a 1D slab, are depicted in figure 3. The introduction of micro pores in the struts (strut porosity) has no effect on the radiative behaviour for smaller wavelengths as the ceria is opaque given the large imaginary part of the refractive index. For larger wavelengths and smaller imaginary part of the refractive index, an increase in the strut porosity at a constant micropore size or an increase in the micropore sizes at a constant strut porosity, respectively, increases the transmittance and reduces the reflectance due to increased void space and reduced averaged solid extinction coefficient. This effect is most significant at small pore sizes or volume fractions and vanishes at larger pore sizes and volume fractions as the strut becomes optically thick due to the strongly increased scattering.

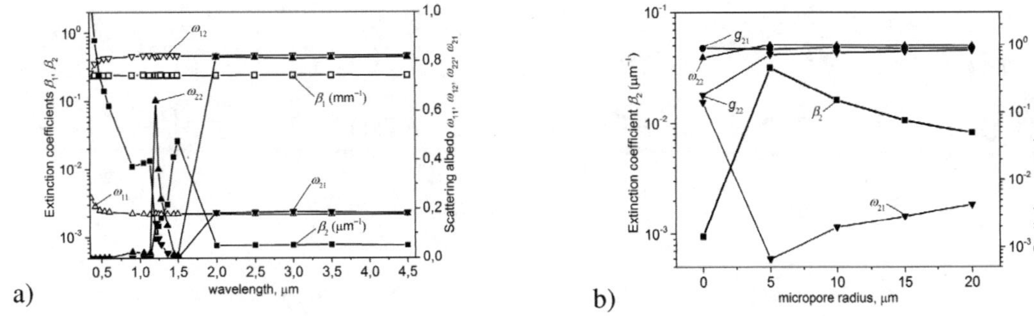

Figure 2. a) Effective radiative properties of the ceria foam with 82% porosity and no microscopic strut porosity, and b) effective radiative properties for the solid phase only for the ceria foam of 82% porosity with additional 10% microporosity in the strut for varying micropore radii.

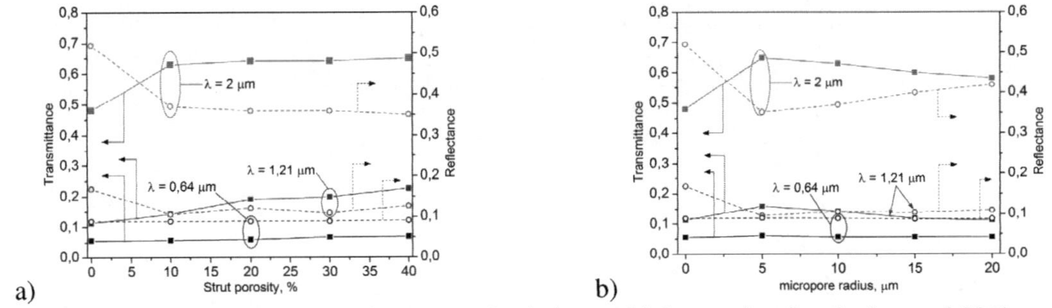

Figure 3. Transmittance and reflectance of a 9.5 mm-thick sample of ceria foam of 82% porosity (a) at constant strut pore size of 20 µm and with increasing strut porosity, and (b) at strut porosity of 10% and with increasing strut pore diameter.

5. Conclusions

The proposed methodology based on the volume-averaged RTEs and Mie theory for the small-scale scatterers allows for the radiative characterization of porous ceramic foams with dual-scale porosity.

References

[1] C Bohren and D Huffman 2004 John Wiley and Sons Inc., New York
[2] WC Chueh, C Falter, M Abbott, D Scipio, P Furler, SM Haile and A Steinfeld 2010 *Science* **330** 1797–1801.
[3] R Coquard, D Baillis and J Randrianalisoa *Int. J. Therm. Sci.*, **50** 1648 – 63.
[4] P Furler, J Scheffe, D Marxer, M Gorbar, A Bonk, U Vogt and A Steinfeld 2014 *PCCP* **16** 10503-11.
[5] S Haussener, P Coray, W Lipinski, P Wyss and A Steinfeld 2010 *J. Heat Trans.* **132** 023305.
[6] W Lipinski, D Keene, S Haussener and J Petrasch 2010 *JQRST* **111** 2474–80.
[7] W Lipinski, J Petrasch and S. Haussener 2010 *JQRST* **111** 253–8.
[8] P Patsalas, S Logothetidis and C Metaxa 2002 *Appl. Phys. Lett.*, **81** 466–468.
[9] A Studart, U Gonzenbach, E Tervoort and L Gauckler 2006 *JACS* **89** 1771–89.
[10] S Suter, A Steinfeld and S Haussener 2014 *Int. J. Heat and Mass Trans.* **78** 688 – 98.
[11] J Randrianalisoa, D Baillis and L Pilon, JOSA A 2006; **23** 1645-56

Iterative and FEM methods to solve the 2–D Radiative Transfer Equation with specular reflexion

David Le Hardy, Yann Favennec, Benoît Rousseau
LTN UMR CNRS 6607,University of Nantes, France
E-mail: david.le-hardy@univ-nantes.fr

Abstract. The present paper deals with iterative algorithms coupled with finite element methods (FEM) to solve the Radiative Transfer Equation (RTE) within semi-transparent heterogenous materials where specular reflexions occur on their boundaries. As our intention is to use such solution for inversion, the forward model should be solved as fastly as possible. This communication compares, in terms of both accuracy and CPU, the Discontinuous Galerkin (DG) method with the Streamline Upwind Petrov-Galerkin (SUPG) method, both being coupled with the Discrete Ordinate Method. Next, several iteratives methods used to accelerate the convergence are compared. These methods are the Gauss-Siedel (GS), the Source-Iteration (SI) and the Successive Over-Relaxation (SOR) methods.

1. Introduction

The studied forward model commonly used to model the propagation of thermal radiation within heterogeneous semi-transparent media is based on the so-called "Radiative Transfer Equation" (with the acronym RTE). Solving directly the RTE constitues one of the main task of the inverse methods aiming to accurately retrieve the volumetric radiative properties from experimental measurements. In the field of material science, these measurements (reflectance, transmittance) are today easily performed by Fourier Transform Infrared Spectrometer on centimetric samples with a 1D plan parallel geometry. Specific set-up adpated to spectrometers allows the simulatenous investigation of the spectral and angular dependances of the radiative properties at a given temperature. The solution of the RTE being very time consuming, for both forward and backward integrations, specific numerical tools have been developped to enhance the speed without altering the quality of the solution. At room temperature, the RTE problem consists in searching the radiance $L(\boldsymbol{x}, \boldsymbol{s})$ satisfying:

$$(\boldsymbol{s} \cdot \nabla + \kappa + \sigma_s) L(\boldsymbol{x}, \boldsymbol{s}) - \sigma_s \oint_{\mathcal{S}^{n-1}} L(\boldsymbol{x}, \boldsymbol{s}')\Phi(\boldsymbol{s}, \boldsymbol{s}') \, \mathrm{d}\boldsymbol{s}' = 0 \quad \forall(\boldsymbol{x}, \boldsymbol{s}) \in \mathcal{D} \times \mathcal{S}^{n-1} \tag{1}$$

where $\boldsymbol{s} \in \mathcal{S}^{n-1}$ is the direction of propagation of L at the location $\boldsymbol{x}$ ($\mathcal{S}^{n-1}$ is the unit circle in 2D), κ and σ_s are the homogeneized absorption and scattering coefficients, respectively, and $\Phi(\boldsymbol{s}, \boldsymbol{s}')$ is the probability density scattering phase function. The spectral dependance is omitted for more clarity.

In applications considered here, the sample is illuminated with a collimated beam of radiation from the direction $\boldsymbol{s}_{\text{in}}$ on the surface boundary $\boldsymbol{x} \in \partial\mathcal{D}_0$ as it is the case for bi-directionnal measurements. Moreover, specular boundary conditions are also considered, as in [1]. This leads to consider the boundary condition formulated as:

$$L(\boldsymbol{x},\boldsymbol{s}) = \left[\tilde{L}_{\text{in}} \times \mathbb{1}_{[\boldsymbol{s}=\boldsymbol{s}_{\text{in}}]} + \rho(\boldsymbol{n}\cdot\boldsymbol{s}_{\text{out}})L(\boldsymbol{x},\boldsymbol{s}_{\text{out}})\right] \times \mathbb{1}_{[\boldsymbol{x}\in\partial\mathcal{D}_0]} \tag{2}$$

with $\rho(.)$, the local reflectivity computed on the interface with the Fresnel law. In most situations, the system of equations (1)-(2) cannot be analytically solved due to the integral term $\oint_{\mathcal{S}^{n-1}} \cdot\, \mathrm{d}\boldsymbol{s}'$ and reflexions $\rho L(\cdot)$. To cope with such difficulties, the DOM approximates the integral by a quadrature $\sum_{j=1}^{\boldsymbol{N}_d} \boldsymbol{w}_j$. The RTE (1) becomes the set of semi-discrete system (SDS) :

$$(\boldsymbol{s}_m\cdot\nabla + \kappa + \sigma_s)\,L_m - \sigma_s \sum_{j=1}^{\boldsymbol{N}_d} \boldsymbol{w}_j L_j \Phi_{m,j} = 0 \quad \forall \boldsymbol{x}\in\mathcal{D},\ \forall m\in 1,\cdots \boldsymbol{N}_d \tag{3}$$

where we used $L_m = L(\boldsymbol{x},\boldsymbol{s}_m)$ and $\Phi_{m,j} = \Phi(\boldsymbol{s}_m,\boldsymbol{s}_j)$.

2. Iterative methods

The DOM yields to write down, at the end, a linear system of the form Ax = b. The dimension of x depends on the number of discrete angles N_t and on the number of degrees of freedom of functional spaces built on the chosen space mesh. Iterative methods may be necessary to reduce the memory capacity:

$$\begin{aligned}\boldsymbol{s}_m\cdot\nabla L_m^{N+1} + (\kappa+\sigma_s - w_m\Phi_{m,m}\sigma_s)L_m^{N+1} &= (1-r)\boldsymbol{s}_m\cdot\nabla L_m^N \\ &+ (1-r)(\kappa+\sigma_s - w_m\Phi_{m,m}\sigma_s)L_m^N + r\sigma_s\sum_{j>m}^{\boldsymbol{N}_d} w_j\Phi_{m,j}L_j^N + r\sigma_s\sum_{j<m}^{\boldsymbol{N}_d} w_j\Phi_{m,j}L_j^\alpha\end{aligned} \tag{4}$$

with $L_m^0 = \tilde{L}_m$, $r\in]0,2[$ is the relaxation parameter, and N and α are iteration numbers.

- For $r=1$ and $\alpha=N$, the iterative scheme is the Source-Iteration (SI). This method uses the radiance at the Nth iteration in all directions but m to compute the next iteration L_m^{N+1}. The Source-Iteration method [2] is popular maybe because of its simplicity.
- For $r=1$ and $\alpha=N+1$, the iterative scheme is the Gauss-Siedel (GS). This method uses the radiance at the Nth iteration in all directions $j>m$, and the radiance computed at the curent $(N+1)^{\text{th}}$ iteration for $j<m$ to compute the next iteration L_m^{N+1}. This Gauss-Siedel method has been applied for the same boundary conditions (2) in [3].
- For $r\in]0,2[$ and $\alpha=N+1$, the iterative scheme is the Sucessive Over-Relaxation (SOR). This method uses both the radiance at the Nth iteration for $j\geq m$, and the radiance computed at the current $(N+1)^{\text{th}}$ iteration $j<m$ to compute the next iteration L_m^{N+1}. The SOR method has been applied for non-LTE boundary conditions in [4].

3. Spacial scheme

The Streamline Upwind Petrov-Galerkin (SUPG) scheme [5] uses the lagrangian $\mathbb{P}_k$ basis to build the variational formulation. The mth-equation of the semi-discrete system (3) – in short (SDS$_m$) – is multiplied by the test function $\boldsymbol{s}_m\cdot\nabla v$ before integration on $\mathcal{D}$:

$$\int_{\mathcal{D}} (\text{SDS}_m)\boldsymbol{s}_m\cdot\nabla v \;\mathrm{d}\boldsymbol{x} \tag{5}$$

Discontinuous Galerkin schemes are hybrid methods combining FEM and FVM (finite volume method). The cells are connected by an echange of flux on the boundary on nearby cells. The (SDS$_m$) is multiplied by a test function v and it is integrated on each $\mathcal{C}_i$ before summation:

$$\int_{\mathcal{C}_i} (\text{SDS}_m)v \;\mathrm{d}\boldsymbol{x} \tag{6}$$

4. Results

The simulations are performed both on a 4 cm^2 square and on a 2 cm radius disk. The collimated beam is a gaussian radiance applied in the $s_{in} = (1, 0)$ direction. We used the following parameters: the refractive indices $n_{\text{out}} = 1$, $n_{\text{in}} = 1.4$, $\kappa = 0.4 \text{ cm}^{-1}$ and the Henyey-Greenstein phase fonction with $g = 0.5$. The errors $L_2(\mathcal{D})$ of the schemes are compared for different meshes at first for $\sigma_s = 0 \text{ cm}^{-1}$. Specifically, $\mathbb{P}_1$ and $\mathbb{P}_2$ SUPG is compared with $\mathbb{P}_0$ and $\mathbb{P}_1$ DG. Next, the convergence of the iterative methods of the schemes are compared for $\sigma_s = 3 \text{ cm}^{-1}$.

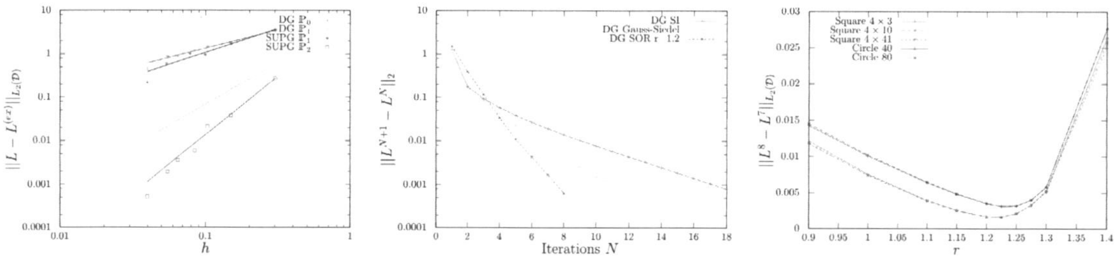

Figure 1. Left: L_2-error between analytical solution and solutions of DG/SUPG schemes as a function of mesh size. Center: convergence error for SI, GS and SOR methods. Right: L_2-error at the 7th iteration for different meshes and two geometries versus the relaxation parameter r.

From Figure 1(a), one can see the error related to the SUPG $\mathbb{P}_2$ scheme is less than for the $\mathbb{P}_0$ and $\mathbb{P}_1$ DG schemes. Moreover, the order of convergence (in space) is greater for the SUPG $\mathbb{P}_2$ scheme. From Figure 1(b), it is seen that the GS method converges faster than the SI method as reported elsewhere in [3]. Moreover, the SOR method, with an appropriate relaxation parameter r, can converge even faster than the GS method. From Figure 1(c), one can see that the choice of the optimal parameter r^* does not depend on the mesh fineness for two given geometries.

5. Conclusion

From these numerical tests, one concludes that (i) the SUPG $\mathbb{P}_2$ scheme is even more appropriate than the DG $\mathbb{P}_1$ scheme in terms of errors, with a smaller number of degrees of freedom, (ii) the iterative SOR method with an appropriate relaxation parameter is a promising extension to more classical SI and GS methods, and (iii) the optimal relaxation parameter being independent of the mesh coarseness, it can be searched on a very coarse mesh.

The direct extension of such work is the 3D implementation where the conclusions presented here should be also drawn. Moreover, the multigrid acceleration [3] will be combined with the SOR method to save again CPU-time significantly. At last, this work will open the possibilities to investigate the volumetric radiative properties of samples with a 3D symmetry.

References

[1] D. Lacroix, N. Berour, P. Boulet, and G. Jeandel, "Transient radiative and conductive heat transfer in non-gray semitransparent two-dimensional media with mixed boundary conditions," *Heat and mass transfer*, vol. 42, no. 4, pp. 322–337, 2006.

[2] E. E. Lewis and W. F. Miller, "Computational methods of neutron transport," 1984.

[3] H. Gao and H. Zhao, "A fast-forward solver of radiative transfer equation," *Transport Theory and Statistical Physics*, vol. 38, no. 3, pp. 149–192, 2009.

[4] L. Léger, L. Chevallier, and F. Paletou, "Fast 2d non-lte radiative modelling of prominences i. numerical methods and benchmark results," *arXiv preprint astro-ph/0703689*, 2007.

[5] P. B. Bochev, M. D. Gunzburger, and J. N. Shadid, "Stability of the supg finite element method for transient advection–diffusion problems," *Computer methods in applied mechanics and engineering*, vol. 193, no. 23, pp. 2301–2323, 2004.

Journal of Physics: Conference Series **676** (2016) 012014 doi:10.1088/1742-6596/676/1/012014

Numerical modelling of transient heat and moisture transport in protective clothing

P Łapka, P Furmański and T S Wisniewski

Institute of Heat Engineering, Warsaw University of Technology, 21/25 Nowowiejska St, 00-665 Warsaw, Poland

E-mail: Piotr.Lapka@itc.pw.edu.pl

Abstract. The paper presents a complex model of heat and mass transfer in a multi-layer protective clothing exposed to a flash fire and interacting with the human skin. The clothing was made of porous fabric layers separated by air gaps. The fabrics contained bound water in the fibres and moist air in the pores. The moist air was also present in the gaps between fabric layers or internal fabric layer and the skin. Three skin sublayers were considered. The model accounted for coupled heat transfer by conduction, thermal radiation and associated with diffusion of water vapour in the clothing layers and air gaps. Heat exchange due to phase transition of the bound water were also included in the model. Complex thermal and mass transfer conditions at internal or external boundaries between fabric layers and air gaps as well as air gap and skin were assumed. Special attention was paid to modelling of thermal radiation which was coming from the fire, penetrated through protective clothing and absorbed by the skin. For the first time non-grey properties as well as optical phenomena at internal or external boundaries between fabric layers and air gaps as well as air gap and skin were accounted for. A series of numerical simulations were carried out and the risk of heat injures was estimated.

1. Introduction

In firefighting and many industries people are exposed to high temperature of environment leading to a risk of serious thermal skin burns. For example in a case of a flash fire, e.g. sudden explosion, the firefighters may be in contact with hot gases or even with flame. Therefore, to prevent thermal injuries they wear the special protective clothing allowing them to safety escape to a secure location. The typical protective clothing is made of several layers of fabric material e.g. outer shell, moisture barrier, thermal insulation and inner layer separated by thin air gaps. Protective performance of garments are directly related to the heat and moisture transfer through such multilayer clothing as well as to heat transfer in the human skin.

Many models for studying heat and moisture transport in multi-layer protective clothing exposed to the flash fire conditions and for assessing possibility of heat injuries are available in a literature. Heat transfer model in a flame resistant single-layer fabric subjected to high heat flux was developed by Torvi and Dale [1]. Their model included thermochemical reaction in the materials as well as absorption of thermal radiation, variable thermal properties and heat transfer across an air space separating the fabric from a test sensor. Their predictions of temperature distributions and times to reach the second-degree burn criterion were in good agreement with experimental measurements. Chitrphiromsri and Kuznetsov [2] presented a complete model of a multilayer clothing, air gap, and skin system exposed to a fire which accounted for heat conduction and thermal radiation as well as moisture transport. Subsequently they analyzed duration of the flash fire exposure on the time after which the second and third degree burns were caused. The model was also used to simulate behavior of different fabric systems and their configurations [3]. Ghazy and Bergstrom [4] developed a model to simulate transient conductive and radiative heat transfer in a protective clothing system consisting of the fire-resistant fabric, the human skin, and the air gap between the fabric and the skin. In next paper [5] they extended the previous model to a system that consisted of three layers of fabric separated by two air gaps, the human skin, and the air gap between the clothing and the skin. In both works they predicted transient heat flux and temperature distribution in the protective clothing, skin and air gap as well as the time to receive skin burn injuries during flash fire heating and subsequent cooling down processes. Zhu and Li [6] developed an interesting numerical model of heat and moisture transfer in a fabric which underwent drying and pyrolysis. The latter was modelled by one-step chemical reaction and lead to changes in fabric thermophysical properties. Their model was validated by experimental data.

New concepts of protecting clothing to improve workers safety were also designed and the respective models describing their thermal behavior were developed. For example Chitrphiromsri et al. [7] proposed an intelligent multi-layer firefighter protective clothing, which during the flash fire exposure absorbed a significant amount of incident heat flux due to evaporation of the injected water and thus limited temperature increase on firefighter's skin. A comprehensive mathematical model of heat and moisture transport was also developed. In turn Hu et al. [8] performed numerical and experimental studies on heat transport through firefighting protective clothing with embedded phase change material (PCM). They simulated different thicknesses and location of PCM in the garment and analyzed possibility of burns occurrence.

In a high temperature environment the main heat transfer mode is thermal radiation which transfers energy from the fire or hot surfaces to the firefighter's clothing and then in conjunction with heat conduction through its textile layers and air gaps to the human skin. The heat exchange process is accompanied by moisture transfer in the fabrics and air gaps which affects thermophisical properties of the clothing system. Therefore, an advanced model describing thermal interaction of a fire with the protecting multi-layer clothing and the conjugated heat and mass transfer across textile layers, air gaps between them and the skin as well as through sublayers of the human skin was developed in this paper. The single fabric layer was assumed to be a hydroscopic porous medium that contained bound water. Additionally the water vapor mixed with the air filled pores in fabric and air gaps. Complex thermal and mass transfer conditions on the internal or external boundaries between fabric layers and air gaps as well as the air gap and the skin were taken into account. Special attention was paid to modeling of thermal radiation coming from the fire, penetrating through the protective clothing and then being absorbed by the skin. For the first time non-grey properties as well as optical phenomena at the internal or external boundaries between fabric layers and air gaps as well as air gap and the skin were considered. Three sublayers skin model was introduced and possibility of heat injures occurrence were also evaluated. A series of numerical simulations for different exposition (heating) times were carried out.

2. Statement of the problem

A schematic diagram of the system under consideration is presented in figure 1. The protective clothing was composed of three fabric layers: outer shell, moist barrier and thermal insulation separated by two narrow air gaps. The external heat source was located on the left hand side while the wide air gap and three sublayers of skin: epidermis, dermis and subcutaneous were on the right hand side. All fabric layers and the skin were assumed to absorb, scatter or reflect and emit thermal radiation and additionally had the same value of refractive index equal to 1.0. All other optical properties were dependent on wavelength and the air was assumed transparent to the thermal radiation. The internal and external boundaries (interfaces) were semitransparent i.e. absorbing and diffusively emitting, reflecting and transmitting thermal radiation except the interfaces between skin sublayers which were assumed transparent. Blood perfusion was included in the skin model. Convection in the clothing and air gaps was neglected. The model assumed one dimensional heat and moisture transfer due to the small thickness of the considered system in comparison to its dimension along the protective clothing and assumed no variation of external thermal conditions in this direction.

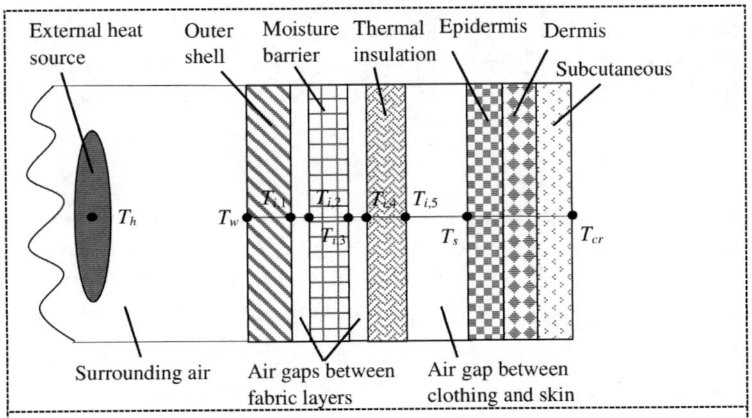

Figure 1. System under consideration.

3. Modelling of the heat transfer in the skin and estimation of degree of burning injuries

The heat transfer model in the skin was based on Pennes model [9] and account for heat transfer between tissue and blood and radiative heat transfer in all skin sublayers [10]. The governing equations in three sublayers were following:

$$(\rho c)_{ep}\frac{\partial T}{\partial t}=\frac{\partial}{\partial x}\left(k_{ep}\frac{\partial T}{\partial x}\right)-\frac{\partial q_{r,ep}}{\partial x} \tag{1}$$

$$(\rho c)_{d}\frac{\partial T}{\partial t}=\frac{\partial}{\partial x}\left(k_{d}\frac{\partial T}{\partial x}\right)+(\rho c)_{b}\,\omega_b\left(T_{cr}-T\right)+q_m-\frac{\partial q_{r,d}}{\partial x} \tag{2}$$

$$(\rho c)_{sc}\frac{\partial T}{\partial t}=\frac{\partial}{\partial x}\left(k_{sc}\frac{\partial T}{\partial x}\right)+(\rho c)_{b}\,\omega_b\left(T_{cr}-T\right)+q_m-\frac{\partial q_{r,sc}}{\partial x} \tag{3}$$

where: subscripts *ep*, *d*, *sc* and *b* denotes epidermis, dermis, subcutaneous and blood, respectively. The symbol ρ describes the density, c is the specific heat, k is the thermal conductivity, ω_b stands for the blood perfusion, q_m is the metabolic heat rate, $T_{cr} = 37°C$ is assumed to be the inner body (core) temperature while q_r denotes radiative heat flux vector. Boundary conditions for these equations were following:

- At the interface between the epidermis and dermis as well as the dermis and subcutaneous continuity of temperature and heat flux were assumed.
- At the right boundary of the system (body interior), see figure 1, fixed body temperature T_{cr} was assumed.
- At left boundary (skin surface), see figure 1, temperature was found from the energy balance in the form:

$$\begin{aligned}&-k_a\left.\frac{\partial T}{\partial x}\right|_a+\int_0^\infty\left(q_{r,\lambda,in,a}-\varepsilon_{\lambda,s}E_{b,\lambda}(T_s)-r_{\lambda,s}q_{r,\lambda,in,a}-t_{\lambda,s}q_{r,\lambda,in,ep}\right)d\lambda+\\&-k_{ep}\left.\frac{\partial T}{\partial x}\right|_s+\int_0^\infty\left(q_{r,\lambda,in,ep}-\varepsilon_{\lambda,s}E_{b,\lambda}(T_s)-r_{\lambda,s}q_{r,\lambda,in,ep}-t_{\lambda,s}q_{r,\lambda,in,a}\right)d\lambda=0\end{aligned} \tag{4}$$

where: the subscripts s and a denote the skin surface and air, respectively, $q_{r,\lambda,in}$ is monochromatic incident radiative heat flux, $E_{b,\lambda}$ is spectral blackbody emissive power, ε, r and t stand for the surface emissivity, reflectivity and transmissivity, respectively and λ denotes the wavelength.

Henriques and Moritz model [11] was applied to estimate a degree of thermal injuries. According to their model the burns occur when temperature at the interface between the epidermis and dermis exceeds $T_{ep/d} = 44°C$. The degree of thermal injury can then be estimated by finding the value of the following integral (Henriques and Moritz integral):

$$\Omega_i = \int_0^t P e^{-\Delta E/BT_i} dt \quad \text{for} \quad T_{ep/d} \geq 44°\text{C} \tag{5}$$

where: t is time, P denotes frequency factor, B describes universal gas constant, ΔE is skin activation energy and T_i stands for interface temperature either between the epidermis and dermis (ep/d) for the first and second degree burns or between the dermis and subcutaneous (d/sc) for the third degree burns. The first degree burns occur for $\Omega_{ep/d} = 0.53$, the second ones for $\Omega_{ep/d} = 1.0$ and the third ones for $\Omega_{d/sc} = 1.0$.

4. Heat transfer in fabric layers and gaps

Heat transfer in the porous fabric layer or in air gap was modeled by the following energy equation which included heat flow by conduction, associated with water vapor diffusion, thermal radiation as well as heat absorbed or released due to phase transition of the bound water to its gaseous (vapor) state:

$$(\rho c)_{ef} \frac{\partial T}{\partial t} = \frac{\partial}{\partial x}\left(k_{ef} \frac{\partial T}{\partial x} + D_{v-a,ef} \frac{\partial \rho_v}{\partial x} c_{p,v} T \right) + \dot{m}_{vbw} \left(\Delta h_{vap} + \Delta h_{abs} \right) - \frac{\partial q_r}{\partial x} \tag{6}$$

The above equation were complemented with additional relations for:

- *Effective heat capacity*:

$$(\rho c)_{ef} = \varepsilon_{bw} \rho_w c_w + \varepsilon_f \rho_f c_f + \varepsilon_g \left(\rho_v c_{p,v} + \rho_a c_{p,a} \right) \tag{7}$$

where dry gas and water vapor densities can be found from formulae:

$$\rho_a = p_a M_a / BT = \varepsilon_a p M_a / BT \tag{8}$$

$$\rho_v = p_v M_v / BT = \varepsilon_v p M_v / BT \tag{9}$$

Here B is the universal gas constant, p, p_a and p_v are total, dry air and water vapor pressures, respectively and M_a and M_v are molecular masses of the dry air and water vapor.

- *Effective thermal conductivity*:

$$k_{ef} = k_g \left\{ \left[1 + \left(\varepsilon_{bw} + \varepsilon_f \right) \right] k_\sigma + \varepsilon_g k_g \right\} \Big/ \left\{ \varepsilon_g k_\sigma + \left[1 + \left(\varepsilon_{bw} + \varepsilon_f \right) \right] k_g \right\} \tag{10}$$

where thermal conductivities of the moist air and fabric as well as dry air are given by the expressions:

$$k_g = \left(k_v \rho_v + k_a \rho_a \right) / \left(\rho_v + \rho_a \right) \tag{11}$$

$$k_\sigma = \left(k_w \rho_w \varepsilon_{bw} + k_f \rho_f \varepsilon_f \right) / \left(\rho_w \varepsilon_{bw} + \rho_f \varepsilon_f \right) \tag{12}$$

$$k_a(T) = \begin{cases} 0.026 + 0.000068(T[\text{K}] - 300) & \text{for} \quad T \leq 700\text{K} \\ 0.053 + 0.000054(T[\text{K}] - 700) & \text{for} \quad T > 700\text{K} \end{cases} \tag{13}$$

- *Effective diffusivity*:

$$D_{v-a,ef} = \begin{cases} D_a \varepsilon_g / \tau & \text{for fabric} \\ D_a & \text{for air gap} \end{cases} \tag{14}$$

where the diffusivity of the water vapor in the air is given by:

$$D_a = 2.23 \cdot 10^{-5} \left(T/273.13 \right)^{1.75} \tag{15}$$

and τ is the tortuosity factor.

In the above formulae: ε_{bw}, ε_f, ε_g, ε_a and ε_v are volumetric fractions of the bound water, dry fabric, moist air, dry air and water vapor, respectively. The specific heats of water, dry fabric, dry air and water vapor are denoted by symbols c_w, c_f, $c_{p,a}$ and $c_{p,v}$, while their thermal conductivities given by k_w, k_f, k_a and k_v. The symbols ρ_w and ρ_f stand for densities of water and dry fabric. The heat of desorption of the bound water to its liquid state and from the liquid state to the gaseous one were assumed to satisfy the following relations:

$$\Delta h_{abs} = 1.95 \cdot 10^5 (1-\phi) \left[(0.2+\phi)^{-1} + (1.05-\phi)^{-1} \right] \tag{16}$$

$$\Delta h_{vap} = 2.792 \cdot 10^6 - 160T[\mathrm{K}] - 3.43T[\mathrm{K}]^2 \tag{17}$$

where the relative humidity is related to the partial pressures of water vapor in the air-vapor mixture by:

$$\phi = p_v / p_s \tag{18}$$

and the water vapor pressure at saturation was calculated from the formula:

$$p_s = 614.3 \exp\left[17.06 (T[\mathrm{K}] - 273.15 / T[\mathrm{K}] - 40.25) \right] \tag{19}$$

The volumetric mass rate of transition of the bound water, present in the fabric, to the gaseous state depends on how far the bound water content is from its equilibrium one:

$$\dot{m}_{vbw} = D_f \rho_f / d_f^2 \left(R_{f,eq} - R_f \right) \tag{20}$$

where the fiber regain and equilibrium fiber regain can be found from expressions:

$$R_f = \varepsilon_{bw} \rho_w / \varepsilon_f \rho_f \tag{21}$$

$$R_{f,eq} = 0.578 R_{f,\phi=0.65} \phi \left[(0.321+\phi)^{-1} + (1.262-\phi)^{-1} \right] \tag{22}$$

Here D_f is the effective diffusivity of bound water in the solid phase (fibers), d_f is the average fiber diameter, $R_{f,\,\phi=0.65}$ is fiber regain at $\phi = 0.65$.

The volume fractions are constrained by the following relation:

$$\varepsilon_{bw} + \varepsilon_f + \varepsilon_g = 1 \tag{23}$$

where: $\varepsilon_g = \varepsilon_a + \varepsilon_v$.

For air gaps the terms in equation (6) corresponding to the phase transition of bound water disappear and properties are calculated for the moist air for which $\varepsilon_g = \varepsilon_a + \varepsilon_v = 1.0$.

On the right hand side of computational domain the boundary condition for equation (6) (temperature of the skin surface T_s) was calculated from equation (4) while on the left hand side external wall temperature T_w (surface of the protective clothing) was found from following balance of energy:

$$-k_{ef}\left.\frac{\partial T}{\partial x}\right|_{w,f}-D_{v-a,ef}\left.\frac{\partial \rho_v}{\partial x}\right|_{w,f}c_{p,v}T_{w,f}+q_{amb}$$
$$+\int_0^\infty\left\{q_{r,\lambda,f,in}-\varepsilon_{\lambda,w}E_{b,\lambda}\left(T_w\right)-r_{\lambda,w}q_{r,\lambda,f,in}-t_{\lambda,w}q_{r,\lambda,amb,in}\right\}d\lambda \tag{24}$$
$$+\int_0^\infty\left\{q_{r,\lambda,amb,in}-\varepsilon_{\lambda,w}E_{b,\lambda}\left(T_w\right)-r_{\lambda,w}q_{r,\lambda,amb,in}-t_{\lambda,w}q_{r,\lambda,f,in}\right\}d\lambda=0$$

The ambient and incident radiative heat fluxes in equation (24) are given by the flowing expressions:

$$q_{amb}=h_{amb}\left(T_{amb}-T_w\right)+h_h\left(T_h-T_w\right)+h_m\left(\rho_{v,amb}-\rho_{v,w}\right)c_{p,v}T_{amb} \tag{25}$$

$$q_{r,\lambda,amb,in}=\varepsilon_{\lambda,h}E_{b,\lambda}\left(T_h\right)+\varepsilon_{\lambda,amb}E_{b,\lambda}\left(T_{amb}\right) \tag{26}$$

where: h_{amb}, h_h and T_{amb}, T_h are heat transfer coefficients and temperatures of the ambient air and hot gases, respectively. The symbol h_m is the ambient air mass transfer coefficient while ε_{amb} and ε_h are the ambient air and hot gases monochromatic emissivities, respectively.

The temperature T_i at the internal interfaces between layers was found from following balance of energy:

$$-k_L\left.\frac{\partial T}{\partial x}\right|_{i,L}-D_{v-a,ef,L}\left.\frac{\partial \rho_v}{\partial x}\right|_{i,L}c_{p,v}T_i-k_R\left.\frac{\partial T}{\partial x}\right|_{i,R}-D_{v-a,ef,R}\left.\frac{\partial \rho_v}{\partial x}\right|_{i,R}c_{p,v}T_i$$
$$+\int_0^\infty\left[q_{r,\lambda,L,in}-\varepsilon_{\lambda,i}E_{b,\lambda}\left(T_i\right)-r_{\lambda,i}q_{r,\lambda,L,in}-t_{\lambda,i}q_{r,\lambda,R,in}\right]d\lambda \tag{27}$$
$$+\int_0^\infty\left[q_{r,\lambda,R,in}-\varepsilon_{\lambda,i}E_{b,\lambda}\left(T_i\right)-r_{\lambda,i}q_{r,\lambda,R,in}-t_{\lambda,i}q_{r,\lambda,L,in}\right]d\lambda=0$$

where the subscripts L and R denote the left and right side of the interface, respectively.

Distribution of temperature in the equilibrium steady state when clothing-skin system is only in contact with the ambient air at temperature T_{amb} were assumed as initial condition for equations (6).

5. Mass transfer in the fabric layers and the air gaps

The mass balance of the water bounded in the fabric, accounting for water sorption and desorption, can be cast in the following form:

$$\frac{\partial \rho_w \varepsilon_{bw}}{\partial t}=\dot{m}_{vbw} \tag{28}$$

Variation of the mass of water vapor filling the pores in the fabric layers and air gaps can be found from the equation:

$$\frac{\partial \rho_v \varepsilon_g}{\partial t}=\frac{\partial}{\partial x}D_{v-a,ef}\frac{\partial \rho_v}{\partial x}-\dot{m}_{vbw} \tag{29}$$

The following boundary conditions were assumed for the equation (29):

- Skin surface was impermeable.
- At the left external wall (clothing surface) density $\rho_{v,w}$ was calculated from the expression:

$$h_m\left(\rho_{v,amb}-\rho_{v,w}\right)=-D_{v-a,ef}\left.\frac{\partial \rho_v}{\partial x}\right|_w \tag{30}$$

- At the internal interfaces density $\rho_{v,i}$ was found from following balance of mass:

$$-D_{v-a,ef,L}\left.\frac{\partial \rho_v}{\partial x}\right|_{i,L} = -D_{v-a,ef,R}\left.\frac{\partial \rho_v}{\partial x}\right|_{i,R} \tag{31}$$

After solving equation (29) the dry air and vapor volume fractions were obtained from the formulae:

$$\varepsilon_a = \frac{p_a}{p}\varepsilon_g \text{ and } \varepsilon_v = \frac{p_v}{p}\varepsilon_g \tag{32}$$

where the respective pressures were calculated from the perfect gas relation and Dalton's Law, i.e.: $p_v = \rho_v BT/M_v$ and $p_a = p - p_v$.

Distributions of the volume fraction of bound water and vapor density in the equilibrium steady state when clothing-skin system is only in contact with the ambient air were assumed as initial conditions for equations (28) and (29).

6. Radiative heat transfer in the fire-clothing-air gap-skin system

Distribution of spectral radiative intensity along path s in a system fire-clothing-air gap-skin is described by the following radiative transfer equation:

$$\frac{dI_\lambda}{ds} = -\left(K_{a,\lambda} + K_{s,\lambda}\right)I_\lambda + K_{a,\lambda}I_{b,\lambda} + \frac{K_{s,\lambda}}{4\pi}\int_{4\pi} I_\lambda \Phi_\lambda\left(\mathbf{s}' \to \mathbf{s}\right)\mathrm{d}\Omega \tag{33}$$

where: I_λ is the spectral radiation intensity, $I_{b,\lambda}$ is the spectral intensity emitted by the blackbody, $K_{a,\lambda}$ and $K_{s,\lambda}$ denote the linear absorption and scattering coefficients, respectively, and Φ_λ is monochromatic scattering phase function. All mechanisms (absorption, scattering and emission) of radiation transport were accounted for in the fabric layers. In the skin, due to it low temperature, the self-emission was neglected. The air contained in the gaps was assumed transparent therefore equation (33) was reduced to $dI_\lambda/ds = 0.0$. The spectral blackbody intensity was calculated applying Planck's Law, i.e.:

$$I_{b,\lambda} = \frac{1}{\pi}E_{b,\lambda}(T) = \frac{1}{\pi}\frac{2hc}{\lambda^5}\left(e^{hc/\lambda kT} - 1\right)^{-1} \tag{34}$$

where: h is Planck constant, c is speed of light in the medium, k is Boltzmann constant.

The boundaries between the surroundings and first fabric layer, between fabric layers and air gaps and between air gap and skin surface were assumed semi-transparent i.e.: emitting, reflecting and transmitting diffusively thermal radiation. Interfaces between the skin sublayers were assumed totally transparent while boundary between the subcutaneous and the body interior was ideally black. Conditions for the radiation intensities at the internal and external boundaries were therefore following:

- External wall (clothing surface) between the surroundings and the first fabric layer:

$$I_{\lambda,w} = \varepsilon_{\lambda,w}I_{b,\lambda}(T_w) + r_{\lambda,w}\frac{q_{r,\lambda,f,in}}{\pi} + t_{\lambda,w}\frac{q_{r,\lambda,amb,in}}{\pi} \quad \text{for } \mathbf{s}\cdot\mathbf{n}_w > 0 \tag{35}$$

where: $\mathbf{n}_w$ is inward normal vector at the external wall.

- Internal interface between the fabric layer and air gap (see figure 2):

$$I_{\lambda,i,L}=\varepsilon_{\lambda,i}I_{b,\lambda}\left(T_i\right)+r_{\lambda,i}\frac{q_{r,\lambda,L,in}}{\pi}+t_{\lambda,i}\frac{q_{r,\lambda,R,in}}{\pi}\quad\text{for}\quad\mathbf{s}\cdot\mathbf{n}_{i,L}>0 \tag{36}$$

$$I_{\lambda,i,R}=\varepsilon_{\lambda,i}I_{b,\lambda}\left(T_i\right)+r_{\lambda,i}\frac{q_{r,\lambda,R,in}}{\pi}+t_{\lambda,i}\frac{q_{r,\lambda,L,in}}{\pi}\quad\text{for}\quad\mathbf{s}\cdot\mathbf{n}_{i,R}>0 \tag{37}$$

where: $\mathbf{n}_{i,L}$ and $\mathbf{n}_{i,R}$ are normal vectors at interface – see figure 2.

- Internal interface between the skin and air gap:

$$I_{\lambda,i,a}=\varepsilon_{\lambda,s}I_{b,\lambda}\left(T_s\right)+r_{\lambda,s}\frac{q_{r,\lambda,a,in}}{\pi}+t_{\lambda,s}\frac{q_{r,\lambda,ep,in}}{\pi}\quad\text{for}\quad\mathbf{s}\cdot\mathbf{n}_{i,a}>0 \tag{38}$$

$$I_{\lambda,i,ep}=\varepsilon_{\lambda,s}I_{b,\lambda}\left(T_s\right)+r_{\lambda,s}\frac{q_{r,\lambda,ep,in}}{\pi}+t_{\lambda,s}\frac{q_{r,\lambda,a,in}}{\pi}\quad\text{for}\quad\mathbf{s}\cdot\mathbf{n}_{i,s}>0 \tag{39}$$

- Interface between the subcutaneous sublayer and body interior:

$$I_{\lambda,cr}=I_{b,\lambda}\left(T_{cr}\right)\quad\text{for}\quad\mathbf{s}\cdot\mathbf{n}_{cr}>0 \tag{40}$$

where: $\mathbf{n}_{cr}$ is inward normal vector at the interface.

The additional relationships needed to close the whole system of equations were following:

- Monochromatic incident heat flux which appears in relationships for thermal and optical boundary conditions:

$$q_{r,\lambda,in}=\int_{\mathbf{s}\cdot\mathbf{n}<0}I_{\lambda}\mathbf{s}\cdot\mathbf{n}d\Omega\quad\text{for}\quad\mathbf{s}\cdot\mathbf{n}<0 \tag{41}$$

- Divergence of radiative heat flux vector which appears in energy equations (1)-(3) and (6) in the skin sublayers and clothing, respectively:

$$\frac{\partial q_r}{\partial x}=\int_0^{\infty}K_{a,\lambda}\left(4\pi I_{b,\lambda}\left(T\right)-\int_{4\pi}Id\Omega\right)d\lambda \tag{42}$$

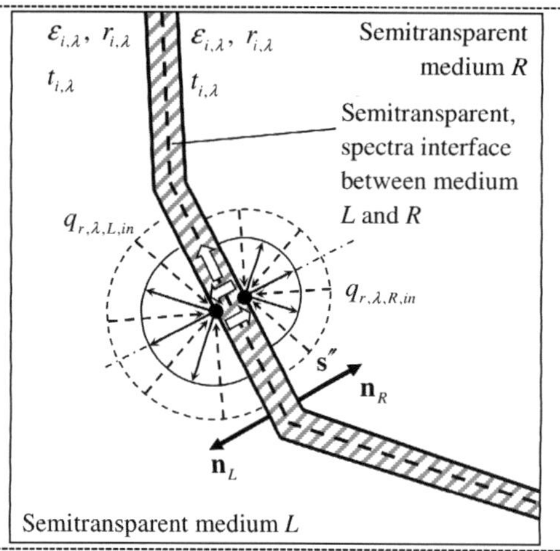

Figure 2. Semitransparent interface

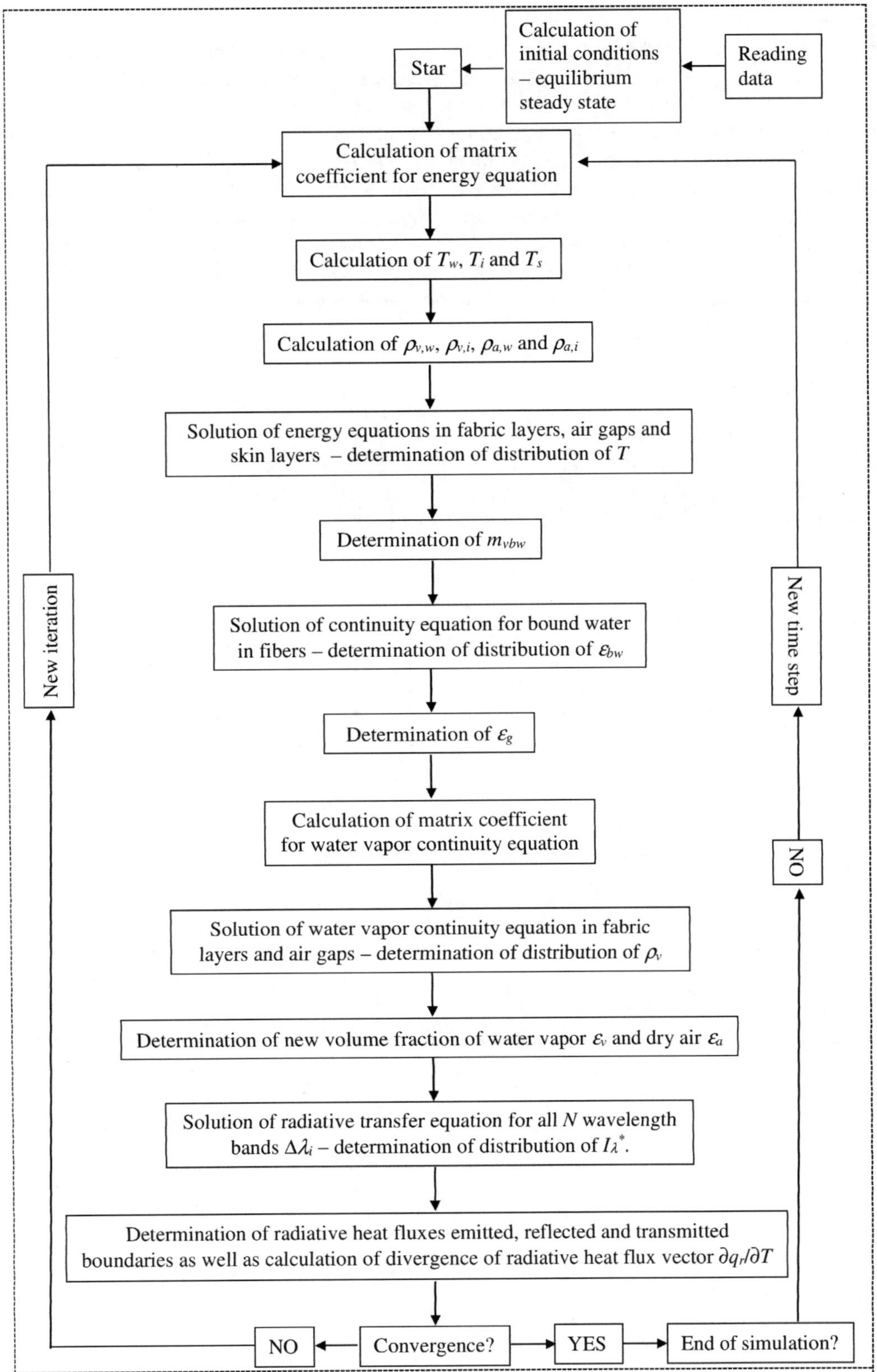

Figure 3. Schematic of computational algorithm.

7. Numerical solution

Energy equations (1)-(3) and (6) in the skin, fabrics layers and air gaps were discretized applying Finite Volume Method on an irregular mesh. Each equation was integrated over infinitesimal distance dx and time interval dt. In the similar way discrete form of continuity equations (28) and (29) for the bound water and water vapour were obtained. Unknown temperature at external and internal interfaces were calculated from conditions (4), (24) and (27) by using Newton-Raphson method. Radiative transfer equation (33) was also discretized by applying Finite Volume Method – the equation was integrated over infinitesimal distance dx, solid angle $d\Omega$ and bandwidth $d\lambda$. Band model was introduced to account for spectral optical properties. All solved equations are coupled, therefore the set of equations obtained was highly nonlinear. The solution was found using the specially developed iterative algorithm which was implemented in the in-house C code. The general schematic of algorithm is presented in figure 3.

8. Results of simulation

Numerical simulations were performed for three layers of the protective clothing. Each layer was separated by a narrow air gap. The widest air gap was between the most inner fabric layer and skin. Configuration of the system is presented in table 1 and the thermophysical properties of fabric layers in table 2. Thermophysical properties of skin and blood are shown in table 3. All data were taken from the available literature [2] and [7]. The optical properties of fabric were assumed as the averaged values presented in [1]-[7] and [12], and are given in table 4. The spectral absorption coefficient (seven bands) for the first layer was assumed using results of measurements presented in [12] – see table 5 and figure 4. Additionally the isotropic scattering with the scattering albedo equal to $\omega = K_a/(K_a+K_s) = 0.5$ were assumed in all fabric layers. The optical properties of the human skin were taken as the averaged ones basing on data cited in [10], [13] and [14]. Constants appearing in equation (5) for determination of the burn degree – Henriques and Moritz integral are given in table 7 and were taken from [3]. Other material properties, boundary and operating parameters were taken from [2] and [7] and had the following values: $\rho_w = 998.2$ kgm^{-3}, $c_w = 4185$ Jkg^{-1}K^{-1}, $k_w = 0.5984$ Wm^{-1}K^{-1}, $k_v = 0.018$ Wm^{-1}K^{-1}, $T_{amb} = 26$°C, $p = 101325$ Pa – constant in the whole system, $\phi_{amb} = 0.68$, $T_h = 1723$°C, $\varepsilon_h = 0.02$, $\varepsilon_{amb} = 1.0$, $h_{amb} = 10$ Wm^{-2}K^{-1}, $h_h = 120$ Wm^{-2}K^{-1} and $h_m = 0.021$ ms^{-1}.

Table 1. Geometrical configuration of the system [2], [7].

First layer (m)	First gap (m)	Second layer (m)	Second gap (m)	Third layer (m)	Third gap (m)	Skin
0.00056	0.0001	0.00073	0.0001	0.00166	0.00635	

Table 2. Thermophysical properties of fabrics [2], [7].

	First layer Kombat 7.5 oz/yd^2	Second layer ComfortZone™	Third layer Aralite™
ρ_f (kgm^{-3})	1384	1295	1380
c_f (Jkg^{-1}K^{-1})	1420	1325	1200
k_f (Wm^{-1}K^{-1})	0.179	0.144	0.130
ε_f	0.334	0.186	0.115
$R_{f,\phi=0.65}$	0.084	0.038	0.045
τ	1.50	1.25	1.00
D_f (m^{2}s^{-1})	$6.0 \cdot 10^{-14}$	$6.0 \cdot 10^{-14}$	$6.0 \cdot 10^{-14}$
d_f (m)	$1.6 \cdot 10^{-5}$	$1.6 \cdot 10^{-5}$	$1.6 \cdot 10^{-5}$

Table 3. Thermophysical properties of the skin and blood [2], [7].

	Epidermis	Dermis	Subcutaneous	Blood
ρ (kgm^{-3})	1200	1200	1000	1060
c (Jkg^{-1}K^{-1})	3598	3222	2760	3770
k (Wm^{-1}K^{-1})	0.255	0.523	0.167	-
L (m)	$8\cdot10^{-5}$	$2\cdot10^{-3}$	$1\cdot10^{-2}$	-
q_m (Wm^{-3})	0	0	0	-
ω_b (s^{-1})	-	-	-	$1.25\cdot10^{-3}$
T_a (°C)	-	-	-	37.0

Table 4. Optical properties of fabrics.

	K_a (m^{-1})	K_s (m^{-1})	Φ	ε	r	t
First layer	Variable	Variable	isotropic	0.661	0.219	0.120
Second layer	3617.2	3617.2	isotropic	0.545	0.268	0.187
Third layer	2025.13	2025.13	isotropic	0.643	0.236	0.121

Table 5. Optical properties of the first fabric layer.

Band no.	1	2	3	4	5	6	7
λ_1 (m)	0	$1.89\cdot10^{-6}$	$2.12\cdot10^{-6}$	$2.70\cdot10^{-6}$	$2.75\cdot10^{-6}$	$3.97\cdot10^{-5}$	$5.77\cdot10^{-5}$
λ_2 (m)	$1.89\cdot10^{-6}$	$2.12\cdot10^{-6}$	$2.70\cdot10^{-6}$	$2.75\cdot10^{-6}$	$3.97\cdot10^{-5}$	$5.77\cdot10^{-5}$	1.0
$K_{a,\lambda}$ (m^{-1})	580.68	720.57	812.85	1000.90	1156.33	1069.31	1183.57
$K_{s,\lambda}$ (m^{-1})	580.68	720.57	812.85	1000.90	1156.33	1069.31	1183.57

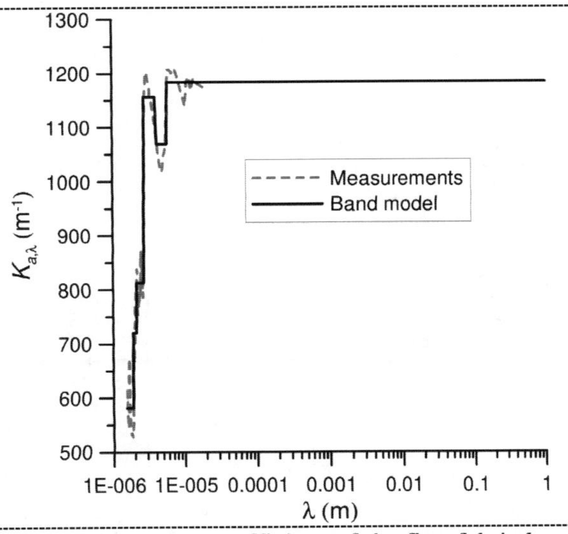

Figure 4. Absorption coefficient of the first fabric layer.

Table 6. Optical properties of the skin.

	K_a (m^{-1})	K_s (m^{-1})	Φ	ε	r	t
Epidermis	200.0	2800.0	isotropic	0.20	0.50	0.30
Dermis	700.0	2000.0	isotropic	-	-	-
Subcutaneous	100.0	1000.0	isotropic	-	-	-

Table 7. Constants appearing in determination of the burn degree [3].

	P (s^{-1})		ΔE (Jkmole^{-1})	
	T_i < 50°C	$T_i \geq$ 50°C	T_i < 50°C	$T_i \geq$ 50°C
First and second degree	2.18 10^{124}	1.82 10^{51}	7.784 10^{8}	3.222 10^{8}
Third degree	4.32 10^{64}	9.39 10^{104}	4.143 10^{8}	6.654 10^{8}

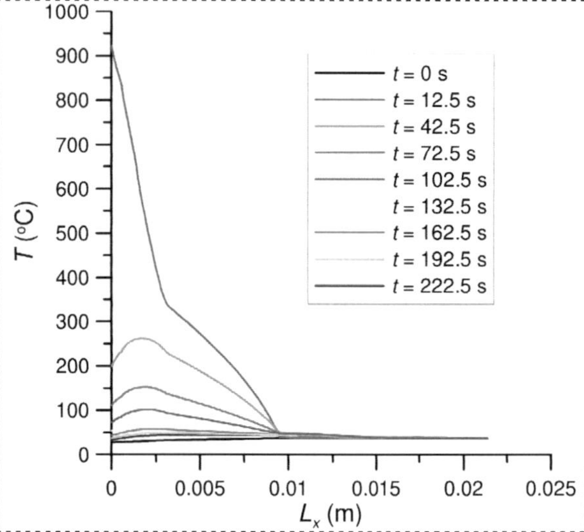

Figure 5. Variation of temperature across the protective clothing, air gaps and skin in time for t_e = 12.5 s.

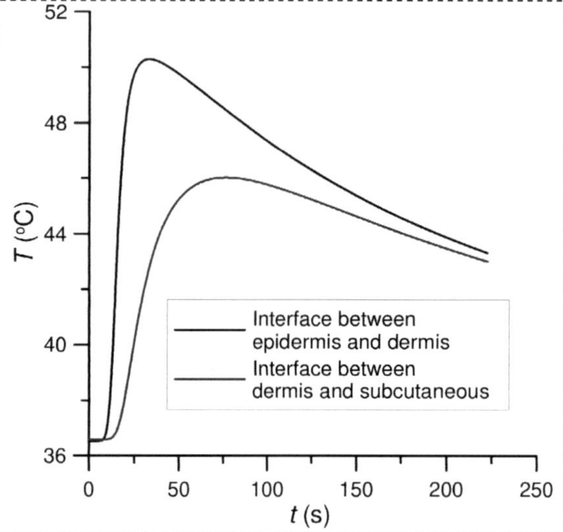

Figure 6. Variation of temperature at interfaces between epidermis and dermis as well as dermis and subcutaneous in time t_e = 12.5 s.

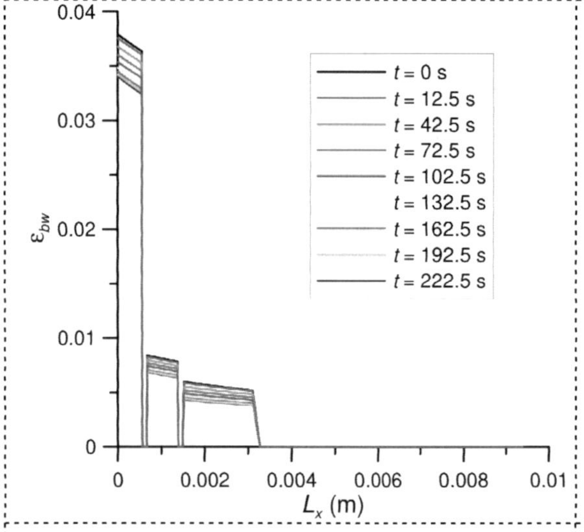

Figure 7. Distribution of moisture content of the bound water across the protective clothing layers in time for t_e = 12.5 s.

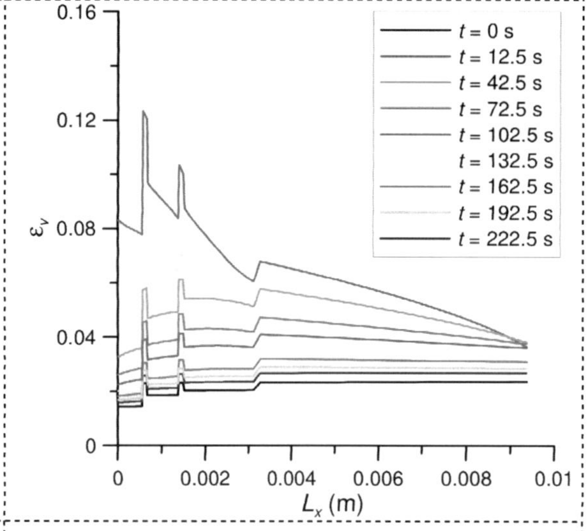

Figure 8. Distribution of water vapour content across the protective clothing layers and air gaps in time for t_e = 12.5 s.

Journal of Physics: Conference Series **676** (2016) 012014 doi:10.1088/1742-6596/676/1/012014

Numerical simulations were performed in the following way. Initially temperature and moisture distributions corresponded to the steady state for interaction of the clothing-skin system with the ambient air at temperature T_{amb}. Subsequently the considered protective clothing was convectively and radiatively heated up by hot gases at temperature T_h. The exposition time t_e varied from 1 to 25 s. Next the protective garment was by 210 s cooled down in the surroundings at temperature T_{amb}.

Figure 5 shows temporary distribution of temperature in the clothing, air gap and the skin for $t_e = 12.5$ s. Figure 6 displays variation of temperature at interface between the epidermis and dermis as well as the dermis and subcutaneous for $t_e = 12.5$ s. During the exposition temperature in the garment significantly increased but the skin temperature did not exceeded 46°C. Figures 7 and 8 present distributions of the volume fraction of bound water and water vapor in the fabric layers and air gaps for different moments of time for $t_e = 12.5$ s. The bound water content in the fibers did not change much in time but the volume fraction of water vapor significantly varied during heating and cooling processes. The high concentration of water vapor can be observed in the air gaps. Distribution of conductive and radiative heat fluxes in the clothing, air gap and the skin after $t_e = 12.5$ s is shown in fig. 9. In the clothing and epidermis heat conduction dominated thermal radiation – conductive heat flux was about two times higher than radiative one. But in the air gap between the most inner fabric layer and the skin radiative heat flux was about three times higher than conductive one. Heat flux associated with moist diffusion in the clothing was in the range of 10 to 100 W/m^2.

Values of Henriques and Moritz integral, calculated from equation 5, are presented in figure 10. The first degree burns appears for exposition time t_e between 11.0 and 11.5 s, the second degree burns for the exposition times higher than $t_e = 12.0$ s while the third degree burns for the exposition times higher than $t_e = 20.0$ s.

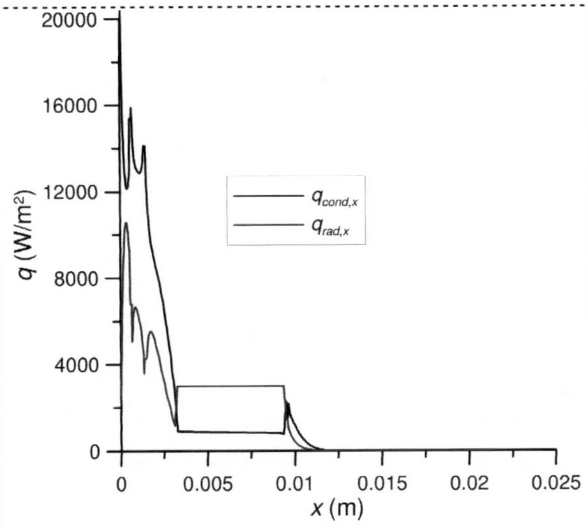

Figure 9. Distribution of conductive and radiative heat fluxes across the protective clothing layers after $t_e = 12.5$ s.

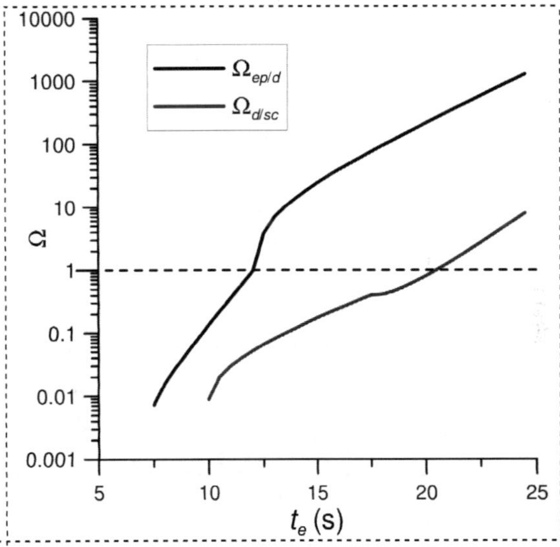

Figure 10. Variation of Henriques and Moritz integral with the exposition time to hot gases.

9. Conclusions

The advanced thermal model of the fire-protective clothing-air gap-skin system was presented in the paper. The model accounted for different modes of heat and mass transfer i.e. heat conduction and thermal radiation in the fabric layers, air gaps and skin, moisture diffusion in the fabrics pores and air gaps as well sorption and desorption of water in textile fibers. Complex thermal and mass transfer conditions on the internal or external boundaries between the fabric layers and air gaps as well as between the air gap and skin were introduced. Special attention was paid to modelling of thermal radiation which was coming from the fire, penetrated through the protective clothing and was absorbed by the skin. For the first time non-grey properties as well as optical phenomena at the internal or external boundaries between the fabric layers and air gaps as well as air gap and skin were accounted for. All governing equations were discretized with Finite Volume Method and an iterative solution algorithm of high non-linear system of equations was elaborated. Simulations were performed for different exposition (heating) times and fixed cooling down period. Distributions of temperatures as well the bound water and water vapor contents were calculated. It was found that on contrary to the bound water the water vapor content significantly varied in time. Share of heat conduction and thermal radiation in total heat transfer were analyzed. Subsequently the Henriques and Moritz integral was calculated allowing the risk of skin burns to be evaluated.

References

[1] Torvi D A and Dale J D 1999 *Fire Technol.* **35** 210-31
[2] Chitrphiromsri P and Kuznetsov A V 2005 *Heat Mass Transfer* **41** 206-15
[3] Song G, Chitrphiromsri P and Ding D 2008 *Int. J. Occup. Saf. Ergon.* **14** 89-106
[4] Ghazy A and Bergstrom D J 2010 *Numer. Heat Transfer* A **58** 702-24
[5] Ghazy A and Bergstrom D J 2012 *Numer. Heat Transfer* A **61** 569-93
[6] Zhu F and Li K 2011 *Fire Technol.* **47** 801-19
[7] Chitrphiromsri P, Kuznetsov A V, Song G and Barker R L 2006 *Numer. Heat Transfer* **A** 427-50
[8] Hu Y, Huang D, Qi Z, He S, Yang H and Zhang H 2013 *Heat Mass Transfer* **49** 567-573
[9] Pennes H H 1948 *J. Appl. Physiol.* **1** 93-122
[10] Bashkatov A N, Genina E A and Tuchin V V (2011) *J. Innov. Opt. Heal. Sci.* **4** 9-38
[11] Henriques Jr F C and Moritz A R (1947) Am. J. Pathol. 23 531-49
[12] Lawson J R, Walton W D, Bryner N P and Amon F K 2005 *Estimates of Thermal Properties for Fire Fighters' Protective Clothing Materials* (Gaithersburg: National Institute of Standards and Technology)
[13] Stroup D W, McLane R A and Twilley W H 2007 *Full Ensemble and Bench Scale Testing of Fire Fighter Protective Clothing* (Gaithersburg: National Institute of Standards and Technology)
[14] Lister T, Wright P A, Chappell P H 2012 *J. Biomed. Opt.* **17** 090901.

Acknowledgments

This work is supported by the National Centre for Research and Development (Poland) under Grant No. O ROB/0011/03/001.

Work performed by Piotr Łapka has been partially supported by the EU in the framework of ESF through the "Didactic Development Program of the Faculty of Power and Aeronautical Engineering of the WUT".

Continuum radiative heat transfer modeling in multi-component anisotropic media in the limit of geometrical optics

W. Lipiński[1], **S. Haussener**[2], **and J. Petrasch**[3]

[1] Research School of Engineering, The Australian National University, Canberra, ACT 2601, Australia

[2] Institute of Mechanical Engineering, Ecole Polytechnique Fédérale de Lausanne, Lausanne 1015, Switzerland

[3] Energy Research Center, Vorarlberg University of Applied Sciences, Dornbirn 6850, Austria

E-mail: `wojciech.lipinski@anu.edu.au`

Abstract. Continuum-scale equations of radiative transfer and corresponding boundary conditions are derived for a multi-component anisotropic medium consisting of components in the range of geometrical optics. The derivations are obtained by employing the volume-averaging theory. This study generalizes the previous derivations obtained for multi-component isotropic media.

1. Introduction

Analysis of radiative heat transfer in heterogeneous media is pertinent to a variety of technological and scientific applications including combustion, solar energy, thermal insulations and space systems, and natural environments such as porous soils, snow, and wood. These media often consist of morphological features with the characteristic size significantly larger than the radiation wavelength of interest and, therefore, optical and radiative analyses can be based on the laws of geometrical optics. The radiative transfer equation (RTE) is typically employed as the theoretical basis for radiative characterization of such media assuming the media are homogeneous and described by a set of radiative properties [1]. The radiative properties—the extinction and scattering coefficients, and the scattering phase function—can be obtained theoretically or based on experimental measurements. A phenomenological extension of the RTE to heterogeneous media is based on the introduction of apparent radiative properties properties, which account for the actual morphology and composition of the multi-component medium. Theoretical and experimental approaches to determination of the apparent radiative characteristics of heterogeneous media are discussed in [1–3]. Volume-averaging approaches directly account for the multi-component composition and complex morphology of the medium. They result in coupled volume-averaged RTEs, one for each component or phase [4–6]. The derivation are based on the assumption that the media and their components are isotropic. The latter assumption limits the applicability of the volume-averaging approaches to multi-component anisotropic media, found in many applications such as described in [7].

Previous pertinent studies on directional radiative characterization of anisotropic heterogeneous media focused on ordered, layered structures [8] and fibrous materials [9, 10]. In this paper, the continuum-scale RTEs for multi-component anisotropic media with components in the limit of geometrical optics are derived along with the associated boundary conditions and radiative property definitions by employing the volume-averaging theory. The theory presented is an extention of the previous derivations for multi-component isotropic media presented in [5,6].

2. Continuum-scale equations of radiative transfer

Consider radiative transfer in a multi-component anisotropic medium. The medium consists of $i = 1, \ldots, M_1$ and $i = M_1 + 1, \ldots, M$ semi-transparent and opaque components, respectively, each of an arbitrary shape. Each component i is adjacent to $j = 1, \ldots, N_{i,1}$ and $j = N_{i,1} + 1, \ldots, N_i$ semi-transparent and opaque components, respectively. The present analysis is subject to the following assumptions: (i) all components are non-polarizing and the state of polarization can be neglected; (ii) all components are at local thermodynamic equilibrium; (iii) characteristic dimensions of all components are much larger than the radiation wavelengths of interest so that laws of geometrical optics are valid in each component; (v) diffraction effects are negligible; (vi) dependent-scattering effects due to inter-component radiative interactions are negligible; (vii) all components are at rest as compared to the speed of light; $(viii)$ radiative transfer in each component and the whole medium is quasi-steady.

Each component i is characterized by the set of the discrete-scale optical and radiative properties: the effective refractive index $n_i(\hat{\mathbf{s}})$, the absorption and scattering coefficients, $\kappa_{\lambda,\mathrm{d},i}(\vec{\mathbf{r}}, \hat{\mathbf{s}})$ and $\sigma_{\lambda,\mathrm{s,d},i}(\vec{\mathbf{r}}, \hat{\mathbf{s}})$, respectively, and the scattering phase function $\Phi_{\lambda,\mathrm{d},i}(\vec{\mathbf{r}}, \hat{\mathbf{s}}_\mathrm{i}, \hat{\mathbf{s}})$. Note that the refractive indices are assumed to be indepdenent of wavelength and position for all components. Furthermore, each component i is characterized by its temperature $T_i(\vec{\mathbf{r}})$, i.e. the components are allowed to be at thermal non-equilibrium with respect to each other.

The quasi-steady discrete-scale intensity in each component i can be determined by solving the corresponding quasi-steady discrete-scale RTEs [1]:

$$\hat{\mathbf{s}} \cdot \nabla_{\vec{\mathbf{r}}} L_i(\vec{\mathbf{r}}, \hat{\mathbf{s}}) = -\beta_{\mathrm{d},i}(\vec{\mathbf{r}}, \hat{\mathbf{s}}) L_i(\vec{\mathbf{r}}, \hat{\mathbf{s}}) + \kappa_{\mathrm{d},i}(\vec{\mathbf{r}}, \hat{\mathbf{s}}) L_{\mathrm{b},i}(\vec{\mathbf{r}}) + \frac{\sigma_{\mathrm{s,d},i}(\vec{\mathbf{r}}, \hat{\mathbf{s}})}{4\pi} \int_{\Omega_\mathrm{i}=0}^{4\pi} L_i(\vec{\mathbf{r}}, \hat{\mathbf{s}}_\mathrm{i}) \Phi_{\mathrm{d},i}(\vec{\mathbf{r}}, \hat{\mathbf{s}}_\mathrm{i}, \hat{\mathbf{s}}) \mathrm{d}\Omega_\mathrm{i}, \quad i = 1, \ldots, M, \tag{1}$$

where the spectral subscript λ has been omitted for brevity. $L_{\mathrm{b},i}$ is the spectral blackbody intensity inside the component i. Equation (1) is subject to the boundary conditions at $A_{ij,\hat{\mathbf{s}}\cdot\hat{\mathbf{n}}_{ji}>0}$:

$$L_i(\vec{\mathbf{r}}_{ij}, \hat{\mathbf{s}}) = \int_{\Omega_{\hat{\mathbf{s}}_\mathrm{i}} \cdot \hat{\mathbf{n}}_{ji}>0} \tau_{ji}^{''}(\vec{\mathbf{r}}_{ij}, \hat{\mathbf{s}}_\mathrm{i}, \hat{\mathbf{s}}) L_j(\vec{\mathbf{r}}_{ij}, \hat{\mathbf{s}}_\mathrm{i}) \hat{\mathbf{s}}_\mathrm{i} \cdot \hat{\mathbf{n}}_{ji} \mathrm{d}\Omega_\mathrm{i} - \int_{\Omega_{\hat{\mathbf{s}}_\mathrm{i}} \cdot \hat{\mathbf{n}}_{ji}<0} \rho_{ij}^{''}(\vec{\mathbf{r}}_{ij}, \hat{\mathbf{s}}_\mathrm{i}, \hat{\mathbf{s}}) L_i(\vec{\mathbf{r}}_{ij}, \hat{\mathbf{s}}_\mathrm{i}) \hat{\mathbf{s}}_\mathrm{i} \cdot \hat{\mathbf{n}}_{ji} \mathrm{d}\Omega_\mathrm{i}, \quad j = 1, \ldots, N_{i,1}, \tag{2a}$$

$$L_i(\vec{\mathbf{r}}_{ij}, \hat{\mathbf{s}}) = \varepsilon_{ji}^{'}(\vec{\mathbf{r}}_{ij}, \hat{\mathbf{s}}) L_{\mathrm{b},j}(\vec{\mathbf{r}}_{ij}) - \int_{\Omega_{\hat{\mathbf{s}}_\mathrm{i}} \cdot \hat{\mathbf{n}}_{ji}<0} \rho_{ij}^{''}(\vec{\mathbf{r}}_{ij}, \hat{\mathbf{s}}_\mathrm{i}, \hat{\mathbf{s}}) L_i(\vec{\mathbf{r}}_{ij}, \hat{\mathbf{s}}_\mathrm{i}) \hat{\mathbf{s}}_\mathrm{i} \cdot \hat{\mathbf{n}}_{ji} \mathrm{d}\Omega_\mathrm{i}, \quad j = N_{i,1} + 1, \ldots, N_i, \tag{2b}$$

where $\vec{\mathbf{r}}_{ij}$ is a position vector at the interface A_{ij}, and $\varepsilon_{ji}^{'}$ and $L_{\mathrm{b},j}(\vec{\mathbf{r}}_{ij})$ are the directional spectral emissivity of the interface between components j and i and the blackbody intensity emitted by the component j into the component i, respectively. A superficial average is taken for each term in Eq. (1). Applying the spatial averaging theorem (SAT) [11],

$$\langle \nabla_{\vec{\mathbf{r}}} L_i(\vec{\mathbf{r}}, \hat{\mathbf{s}}) \rangle = \nabla_{\vec{\mathbf{x}}} I_i(\vec{\mathbf{x}}, \hat{\mathbf{s}}) - \frac{1}{V(\vec{\mathbf{x}})} \sum_{j=1}^{N_i} \int_{A_{ij}} L_i(\vec{\mathbf{r}}, \hat{\mathbf{s}}) \hat{\mathbf{n}}_{ji} \mathrm{d}A, \qquad I_i(\vec{\mathbf{x}}, \hat{\mathbf{s}}) \stackrel{\mathrm{def}}{=} \langle L_i(\vec{\mathbf{r}}, \hat{\mathbf{s}}) \rangle \tag{3}$$

to the resulting term containing the intensity gradient, and following the further derivation steps as elaborated for the case of isotropic multi-component media [6] leads to

$$\hat{\mathbf{s}}\cdot\nabla_{\vec{\mathbf{x}}}I_i(\vec{\mathbf{x}},\hat{\mathbf{s}}) = -\left[\beta_{\mathrm{d},i}(\vec{\mathbf{x}},\hat{\mathbf{s}}) + \sum_{j=N_{i,1}+1}^{N_i}\kappa_{ij}(\vec{\mathbf{x}},\hat{\mathbf{s}}) + \sum_{j=1}^{N_{i,1}}\sigma_{\mathrm{s,t},ij}(\vec{\mathbf{x}},\hat{\mathbf{s}}) + \sum_{j=1}^{N_i}\sigma_{\mathrm{s,r},ij}(\vec{\mathbf{x}},\hat{\mathbf{s}})\right] I_i(\vec{\mathbf{x}},\hat{\mathbf{s}})$$

$$+\kappa_{\mathrm{d},i}(\vec{\mathbf{x}},\hat{\mathbf{s}})\,I_{\mathrm{b},i}(\vec{\mathbf{x}},\hat{\mathbf{s}}) + \sum_{j=N_{i,1}+1}^{N_i}\kappa_{ji}I_{\mathrm{b},j}(\vec{\mathbf{x}},\hat{\mathbf{s}}) + \frac{\sigma_{\mathrm{s,d},i}(\vec{\mathbf{x}},\hat{\mathbf{s}})}{4\pi}\int_{\Omega_i=0}^{4\pi} I_i(\vec{\mathbf{x}},\hat{\mathbf{s}}_{\mathrm{i}})\,\Phi_{\mathrm{d},i}(\vec{\mathbf{x}},\hat{\mathbf{s}}_{\mathrm{i}},\hat{\mathbf{s}})\,\mathrm{d}\Omega_{\mathrm{i}}$$

$$+\sum_{j=1}^{N_{i,1}}\frac{\sigma_{\mathrm{s,t},ji}(\vec{\mathbf{x}},\hat{\mathbf{s}})}{4\pi}\int_{\Omega_i=0}^{4\pi} I_j(\vec{\mathbf{x}},\hat{\mathbf{s}}_{\mathrm{i}})\,\Phi_{\mathrm{t},ji}(\vec{\mathbf{x}},\hat{\mathbf{s}}_{\mathrm{i}},\hat{\mathbf{s}})\,\mathrm{d}\Omega_{\mathrm{i}} + \sum_{j=1}^{N_i}\frac{\sigma_{\mathrm{s,r},ij}(\vec{\mathbf{x}},\hat{\mathbf{s}})}{4\pi}\int_{\Omega_i=0}^{4\pi} I_i(\vec{\mathbf{x}},\hat{\mathbf{s}}_{\mathrm{i}})\,\Phi_{\mathrm{r},ij}(\vec{\mathbf{x}},\hat{\mathbf{s}}_{\mathrm{i}},\hat{\mathbf{s}})\,\mathrm{d}\Omega_{\mathrm{i}},$$

$$i = 1,\ldots,M_1. \qquad (4)$$

The continuum-scale absorption and scattering coefficients, and the scattering phase functions, associated with the superficial average intensities I_i and $I_{\mathrm{b},j}$ appearing in Eq. (1), are defined as

$$\kappa_{ij}(\vec{\mathbf{x}},\hat{\mathbf{s}}) \stackrel{\mathrm{def}}{=} -\frac{\displaystyle\int_{A_{ij,\hat{\mathbf{s}}\cdot\hat{\mathbf{n}}_{ji}<0}} \alpha'_{ij}(\vec{\mathbf{r}}_{ij},\hat{\mathbf{s}})\,L_i(\vec{\mathbf{r}}_{ij},\hat{\mathbf{s}})\,\hat{\mathbf{s}}\cdot\hat{\mathbf{n}}_{ji}\,\mathrm{d}A}{I_i(\vec{\mathbf{x}},\hat{\mathbf{s}})\,V(\vec{\mathbf{x}})}, \qquad (5a)$$

$$\kappa_{ji}(\vec{\mathbf{x}},\hat{\mathbf{s}}) \stackrel{\mathrm{def}}{=} \frac{\displaystyle\int_{A_{ij,\hat{\mathbf{s}}\cdot\hat{\mathbf{n}}_{ji}>0}} \varepsilon'_{ji}(\vec{\mathbf{r}}_{ij},\hat{\mathbf{s}})\,L_{\mathrm{b},j}(\vec{\mathbf{r}}_{ij},\hat{\mathbf{s}})\,\hat{\mathbf{s}}\cdot\hat{\mathbf{n}}_{ji}\,\mathrm{d}A}{I_{\mathrm{b},j}(\vec{\mathbf{x}},\hat{\mathbf{s}})\,V(\vec{\mathbf{x}})}, \qquad (5b)$$

$$\sigma_{\mathrm{s,r},ij}(\vec{\mathbf{x}},\hat{\mathbf{s}}) \stackrel{\mathrm{def}}{=} -\frac{\displaystyle\int_{A_{ij,\hat{\mathbf{s}}\cdot\hat{\mathbf{n}}_{ji}<0}} \rho'^{\frown}_{ij}(\vec{\mathbf{r}}_{ij},\hat{\mathbf{s}})\,L_i(\vec{\mathbf{r}}_{ij},\hat{\mathbf{s}})\,\hat{\mathbf{s}}\cdot\hat{\mathbf{n}}_{ji}\,\mathrm{d}A}{I_i(\vec{\mathbf{x}},\hat{\mathbf{s}})\,V(\vec{\mathbf{x}})}, \qquad (5c)$$

$$\sigma_{\mathrm{s,t},ij}(\vec{\mathbf{x}},\hat{\mathbf{s}}) \stackrel{\mathrm{def}}{=} -\frac{\displaystyle\int_{A_{ij,\hat{\mathbf{s}}\cdot\hat{\mathbf{n}}_{ji}<0}} \tau'^{\frown}_{ij}(\vec{\mathbf{r}}_{ij},\hat{\mathbf{s}})\,L_i(\vec{\mathbf{r}}_{ij},\hat{\mathbf{s}})\,\hat{\mathbf{s}}\cdot\hat{\mathbf{n}}_{ji}\,\mathrm{d}A}{I_i(\vec{\mathbf{x}},\hat{\mathbf{s}})\,V(\vec{\mathbf{x}})}, \qquad (5d)$$

$$\Phi_{\mathrm{r},ij}(\vec{\mathbf{x}},\hat{\mathbf{s}}_{\mathrm{i}},\hat{\mathbf{s}}) \stackrel{\mathrm{def}}{=} -\frac{\displaystyle\int_{A_{ij,\hat{\mathbf{s}}\cdot\hat{\mathbf{n}}_{ji}>0}} \rho''_{ij}(\vec{\mathbf{r}}_{ij},\hat{\mathbf{s}}_{\mathrm{i}},\hat{\mathbf{s}})\,L_i(\vec{\mathbf{r}}_{ij},\hat{\mathbf{s}}_{\mathrm{i}})\,\hat{\mathbf{s}}_{\mathrm{i}}\cdot\hat{\mathbf{n}}_{ji}\,\hat{\mathbf{s}}\cdot\hat{\mathbf{n}}_{ji}\,\mathrm{d}A}{(4\pi)^{-1}\,\sigma_{\mathrm{s,r},ij}I_i(\vec{\mathbf{x}},\hat{\mathbf{s}})\,V(\vec{\mathbf{x}})}, \quad \hat{\mathbf{s}}_{\mathrm{i}}\cdot\hat{\mathbf{n}}_{ji} < 0, \qquad (5e)$$

$$\Phi_{\mathrm{t},ji}(\vec{\mathbf{x}},\hat{\mathbf{s}}_{\mathrm{i}},\hat{\mathbf{s}}) \stackrel{\mathrm{def}}{=} \frac{\displaystyle\int_{A_{ij,\hat{\mathbf{s}}\cdot\hat{\mathbf{n}}_{ji}>0}} \tau''_{ji}(\vec{\mathbf{r}}_{ij},\hat{\mathbf{s}}_{\mathrm{i}},\hat{\mathbf{s}})\,L_j(\vec{\mathbf{r}}_{ij},\hat{\mathbf{s}}_{\mathrm{i}})\,\hat{\mathbf{s}}_{\mathrm{i}}\cdot\hat{\mathbf{n}}_{ji}\,\hat{\mathbf{s}}\cdot\hat{\mathbf{n}}_{ji}\,\mathrm{d}A}{(4\pi)^{-1}\,\sigma_{\mathrm{s,t},ji}I_j(\vec{\mathbf{x}},\hat{\mathbf{s}})\,V(\vec{\mathbf{x}})}, \quad \hat{\mathbf{s}}_{\mathrm{i}}\cdot\hat{\mathbf{n}}_{ji} > 0. \qquad (5f)$$

The set of Eqs. (4) presents generalization of Eqs. (18) and (27) in [5], and Eqs. (33) in [6] for a multi-component anisotropic medium consisting of any number of semi-transparent and opaque components. The definitions of continuum-scale radiative properties, Eqs. (5a)–(5f) provide the mathematical basis for development of numerical techniques for determination of continuum-scale radiative properties utilizing the exact geometry of multi-component media. They require the knowledge of the complete actual and blackbody discrete-scale radiative intensity fields in each component obtained for a selected model problem.

3. Continuum-scale boundary conditions

Equations (4) are subject to boundary conditions at the wall-medium interface at $\hat{\mathbf{s}} \cdot \hat{\mathbf{n}}_{\mathrm{w}} > 0$, where $\hat{\mathbf{n}}_{\mathrm{w}}$ is a unit normal vector pointing from the wall into the medium. The wall is assumed to consist of only a single component that can be either semi-transparent or opaque. The discrete-scale boundary conditions at the boundary of the multi-component medium are formulated analogously to the boundary conditions (2a) and (2b). They read for the semi-transparent and opaque walls, respectively:

$$L_i(\vec{\mathbf{r}}_{i\mathrm{w}}, \hat{\mathbf{s}}) = \int\limits_{\Omega_{\hat{\mathbf{s}}_\mathrm{i}} \cdot \hat{\mathbf{n}}_{\mathrm{w}i} > 0} \tau''_{\mathrm{w}i}(\vec{\mathbf{r}}_{i\mathrm{w}}, \hat{\mathbf{s}}_\mathrm{i}, \hat{\mathbf{s}}) L_\mathrm{w}(\vec{\mathbf{r}}_{i\mathrm{w}}, \hat{\mathbf{s}}_\mathrm{i}) \hat{\mathbf{s}}_\mathrm{i} \cdot \hat{\mathbf{n}}_{\mathrm{w}i} \,\mathrm{d}\Omega_\mathrm{i} - \int\limits_{\Omega_{\hat{\mathbf{s}}_\mathrm{i}} \cdot \hat{\mathbf{n}}_{\mathrm{w}i} < 0} \rho''_{i\mathrm{w}}(\vec{\mathbf{r}}_{i\mathrm{w}}, \hat{\mathbf{s}}_\mathrm{i}, \hat{\mathbf{s}}) L_i(\vec{\mathbf{r}}_{i\mathrm{w}}, \hat{\mathbf{s}}_\mathrm{i}) \hat{\mathbf{s}}_\mathrm{i} \cdot \hat{\mathbf{n}}_{\mathrm{w}i} \,\mathrm{d}\Omega_\mathrm{i}, \quad (6a)$$

$$L_i(\vec{\mathbf{r}}_{i\mathrm{w}}, \hat{\mathbf{s}}) = \varepsilon'_{\mathrm{w}i}(\vec{\mathbf{r}}_{i\mathrm{w}}, \hat{\mathbf{s}}) L_{\mathrm{b,w}}(\vec{\mathbf{r}}_{i\mathrm{w}}) - \int\limits_{\Omega_{\hat{\mathbf{s}}_\mathrm{i}} \cdot \hat{\mathbf{n}}_{\mathrm{w}i} < 0} \rho''_{i\mathrm{w}}(\vec{\mathbf{r}}_{i\mathrm{w}}, \hat{\mathbf{s}}_\mathrm{i}, \hat{\mathbf{s}}) L_i(\vec{\mathbf{r}}_{i\mathrm{w}}, \hat{\mathbf{s}}_\mathrm{i}) \hat{\mathbf{s}}_\mathrm{i} \cdot \hat{\mathbf{n}}_{\mathrm{w}i} \,\mathrm{d}\Omega_\mathrm{i}, \quad (6b)$$

The variation of the discrete-scale radiative properties and the curvature of the wall–medium interface are assumed to be negligible over the interface area associated with the averaging volume V adjacent to said boundary, $A_{i\mathrm{w}, \hat{\mathbf{s}} \cdot \hat{\mathbf{n}}_{\mathrm{w}i} > 0}$. The continuum-scale boundary conditions are obtained by surface averaging of the boundary intensity $L_i(\vec{\mathbf{r}}_{i\mathrm{w}}, \hat{\mathbf{s}})$ over the portion of the wall-medium interface inside the averaging volume V adjacent to the wall, A_w,

$$I_i(\vec{\mathbf{x}}_\mathrm{w}, \hat{\mathbf{s}}) = \int\limits_{\Omega_{\hat{\mathbf{s}}_\mathrm{i}} \cdot \hat{\mathbf{n}}_{\mathrm{w}i} > 0} \tau''_{\mathrm{w}i}(\vec{\mathbf{x}}_\mathrm{w}, \hat{\mathbf{s}}_\mathrm{i}, \hat{\mathbf{s}}) I_{\mathrm{w}i}(\vec{\mathbf{x}}_\mathrm{w}, \hat{\mathbf{s}}_\mathrm{i}) \hat{\mathbf{s}}_\mathrm{i} \cdot \hat{\mathbf{n}}_{\mathrm{w}i} \,\mathrm{d}\Omega_\mathrm{i} - \int\limits_{\Omega_{\hat{\mathbf{s}}_\mathrm{i}} \cdot \hat{\mathbf{n}}_{\mathrm{w}i} < 0} \rho''_{i\mathrm{w}}(\vec{\mathbf{x}}_\mathrm{w}, \hat{\mathbf{s}}_\mathrm{i}, \hat{\mathbf{s}}) I_i(\vec{\mathbf{x}}_\mathrm{w}, \hat{\mathbf{s}}_\mathrm{i}) \hat{\mathbf{s}}_\mathrm{i} \cdot \hat{\mathbf{n}}_{\mathrm{w}i} \,\mathrm{d}\Omega_\mathrm{i}, \quad (7a)$$

$$I_i(\vec{\mathbf{x}}_\mathrm{w}, \hat{\mathbf{s}}) = \varepsilon'_{\mathrm{w}i}(\vec{\mathbf{x}}_\mathrm{w}, \hat{\mathbf{s}}) I_{\mathrm{b,w}i}(\vec{\mathbf{x}}_\mathrm{w}, \hat{\mathbf{s}}) - \int\limits_{\Omega_{\hat{\mathbf{s}}_\mathrm{i}} \cdot \hat{\mathbf{n}}_{\mathrm{w}i} < 0} \rho''_{i\mathrm{w}}(\vec{\mathbf{x}}_\mathrm{w}, \hat{\mathbf{s}}_\mathrm{i}, \hat{\mathbf{s}}) I_i(\vec{\mathbf{x}}_\mathrm{w}, \hat{\mathbf{s}}_\mathrm{i}) \hat{\mathbf{s}}_\mathrm{i} \cdot \hat{\mathbf{n}}_{\mathrm{w}i} \,\mathrm{d}\Omega_\mathrm{i}. \quad (7b)$$

4. Summary and conclusions

A mathematical model for radiative transfer and characterization of multi-component anisotropic participating media with the individual components in the limit of geometrical optics has been formulated. The model includes the set of governing continuum-scale radiative transfer equations, and the associated boundary conditions and radiative property definitions. The formulation presented is a generalization of the previous formulation for multi-component isotropic media with components in the limit of geometrical optics. The model presented finds application in radiative transfer analysis and characterization of a broad range of multi-component anisotropic media encountered in engineered and natural systems.

References

[1] Modest M 2013 *Radiative Heat Transfer* 3rd ed (San Diego: Academic Press)
[2] Baillis D and Sacadura J F 2000 *Journal of Quantitative Spectroscopy and Radiative Transfer* **67** 327–363
[3] Tancrez M and Taine J 2004 *International Journal of Heat and Mass Transfer* **47** 373–383
[4] Gusarov A V 2008 *Physical Review B* **77** 144201
[5] Lipiński W, Petrasch J and Haussener S 2010 *Journal of Quantitative Spectroscopy and Radiative Transfer* **111** 253–258
[6] Lipiński W, Keene D, Haussener S and Petrasch J 2010 *Journal of Quantitative Spectroscopy and Radiative Transfer* **111** 2474–2480
[7] Haussener S and Steinfeld A 2012 *Materials* **5** 192–209
[8] Gusarov A 2013 *Journal of Quantitative Spectroscopy and Radiative Transfer* **116** 156–168
[9] Coquard R and Baillis D 2006 *Journal of Heat Transfer* **128** 1022–1030
[10] Bellet F, Chalopin E, Fichot F, Iacona E and Taine J 2009 *International Journal of Heat and Mass Transfer* **52** 1544–1551
[11] Whitaker S 1999 *The Method of Volume Averaging* (*Theory and Applications of Transport in Porous Media* vol 13) (Dordrecht: Kluwer Academic Publishers)

Radiative modelling by the zonal method and WSGG model in inhomogeneous axisymmetric cylindrical enclosure

Rachid Méchi[1], **Habib Farhat, Rachid Said**
Lab. Studies of Ionized and Reactive Media
Ibn Eljazzar Street 5019 Monastir, Tunisia
E-mail:
Rachid.mechi@isigk.rnu.tn

Abstract
Nongray radiation calculations are carried out for a case problem available in the literature. The problem is a non-isothermal and inhomogeneous CO_2-H_2O-N_2 gas mixture confined within an axisymmetric cylindrical furnace. The numerical procedure is based on the zonal method associated with the weighted sum of gray gases (WSGG) model. The effect of the wall emissivity on the heat flux losses is discussed. It is shown that this property affects strongly the furnace efficiency and that the most important heat fluxes are those leaving through the circumferential boundary. The numerical procedure adopted in this work is found to be effective and may be relied on to simulate coupled turbulent combustion-radiation in fired furnaces.

1. Introduction

Radiation thermal process occurs in many industrial applications, mainly in fired furnaces and boilers. In such systems, resolving the radiative transfer equation (RTE) is necessary to predict the spatial temperature and heat flux distributions as well as the radiative wall heat losses. It is interesting to note that with a comprehensive mathematical model and fairly realistic gas radiative properties model, satisfactory and relatively fast numerical predictions may be obtained. An exhaustive numerical investigation [1] was achieved in a 3-D cylindrical complex geometry. The author coupled a computational fluid dynamics (CFD) code with the radiation-turbulence-chemistry and combustion interactions. Recently, the zone method and the WSGG model both apply more and more to two and three-dimensional complex geometries [2-5]. It these references, the predicted temperature and flux fields are satisfactory as compared with data obtained from experiments conducted on different fired plants.

In the present work, the zonal method formulated in terms of interchange areas associated with a spatial absorption coefficient distribution due to Murty et al's model [6], were applied to deal with thermal radiation transfer within a cylindrical axisymmetric enclosure. The medium is an inhomogeneous CO_2-H_2O-N_2 gas mixture with 10%, 20% and 70% respectively. This case problem was already investigated particularly by Moghari et al [2]. The effectiveness of the numerical procedure adopted is assessed. The internal boundary emissivity effect on the wall flux is discussed as it plays an important role in the heat exchange process which in turn affects the furnace efficiency.

[1] To whom any correspondence should be addressed.

2. Mathematical formulation

In the standard zone method, the enclosure is divided into gas and surface zones which are small enough to justify the assumption that the radiative properties are uniform in each individual zone. First, the direct exchange areas (DEAs) are numerically estimated for each gray gas then used to evaluate the total exchange areas (TEAs) by solving matrix relations [3, 7, 8]. The former needs only to be assessed once for each single gray gas before solving the governing energy balance equations overall the surface and volume zones. The net radiative heat flux on a surface zone A_i may be expressed as [8]

$$q_i = \sum_{j=1}^{\Gamma_s} \overrightarrow{S_j S_i}\, E^{\circ}_{s,j} + \sum_{k=1}^{\Gamma_g} \overrightarrow{G_k S_i}\, E^{\circ}_{g,k} - A_i \varepsilon_i E^{\circ}_{s,i} \tag{2}$$

where $E_{s,j}$ and $E_{g,k}$ are the blackbody emissive powers of surface zone j and gas volume k, respectively. $\overrightarrow{S_i S_j}$ and $\overrightarrow{G_i S_j}$ represent the directed flux areas (DFAs) for surface-to-surface and gas-to-surface radiative exchange, respectively. The WSGG model has been initially developed by Hottel and Sarofim within the framework of the zonal method [9]. It consists in substituting a set of several virtual gas components for the real gas. Each gray gas is assigned a constant absorption coefficient, and thus the total gas or gas-soot mixture emissivity is

$$\varepsilon_g(T) = \sum_n a_{g,n}(T)[1 - e^{K_{g,n}L}] \tag{3}$$

DEAs are calculated for an axisymmetric cylindrical geometry confining an inhomogeneous CO_2-H_2O-N_2 gas mixture. A bilinear interpolation scheme is thus used to evaluate the absorption coefficient of the inhomogeneous medium at any point. Integration procedure is based on the trapezoidal scheme together with the Richardson acceleration algorithm. A three-dimensional integral is calculated for each zone pair by integrating over the path joining their midpoints and the full 2π azimuthal interval. Axisymmetry and reciprocity of DEAs make it possible to reduce considerably the computation requirements.

3. Analysis of the results

To check the validity of the proposed numerical procedure it is considered an axisymmetric cylindrical enclosure (radius: 1 m, height=4 m) filled with a CO_2-H_2O-N_2 gas mixture at atmospheric pressure. The temperature distribution is due to Coelho [10] taken later by Moghari et al [2]. The absorption coefficient field is calculated as in [6].

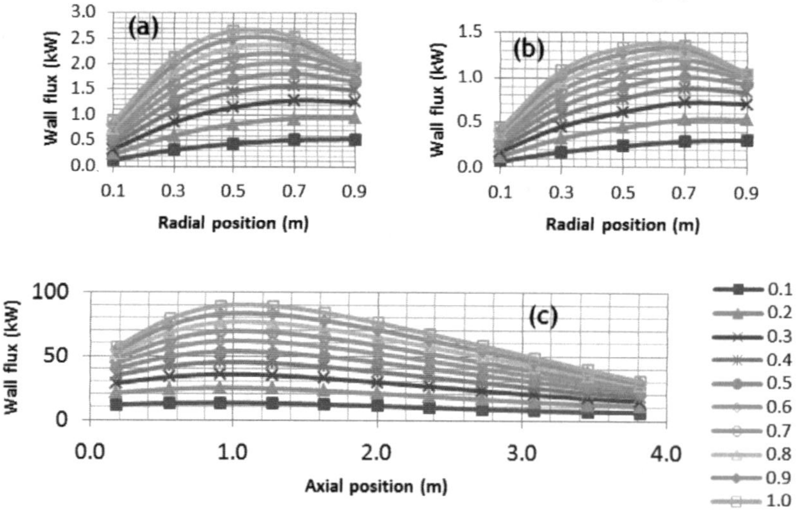

Fig. 2 Wall emissivity effect on the wall heat flux through the (a) lower base (b) upper base and (c) lateral surface.

Figure 2 depicts the net radiative wall flux on both bases and the circumferential surface of the cylindrical system. It is shown that the wall flux losses increase with the internal boundary emissivity on all the surface zones. This flux is found to be zero for totally reflecting internal surface. The flux increases with the radial position starting from zero due to axisymmetry of the studied problem, tends to be uniform close to the centre then decreases. It is worth noting that the flux on a given surface zone, lying on the lower base, is almost the double of its value on the corresponding surface zone of the upper base located at the same radial coordinate. This is foreseeable as the enclosure is a bottom fired system (Fig. 2 a-b). The flux through the circumferential boundary zones takes excessive values as compared to those calculated on the different surface zones of the two bases as shown in Fig. 2-c. It is usually best to consider small wall emissivities in order to minimize energy losses. Actually, the radiative wall property is subject to changes due to the wall deposits such as ash. This alters the furnace efficiency and thus the enclosure requires a regular cleaning.

4. Conclusion

The zonal method associated with the WSGG model is applied to numerically predict wall heat fluxes in an axisymmetric cylindrical enclosure. The participating medium is an inhomogeneous and non-isotherme CO_2-H_2O-N_2 gas mixture. The absorption coefficient along a path length joining the centers of a zone pair, is obtained by bilinear interpolation. The current results show that the radiative thermal losses are higher in the lateral surface than those occurring in both upper and lower bases. Maximum heat exchange occurred at the centre region of the surrounding boundaries as the system is a bottom fired furnace. The effect of the internal wall emissivity on the fired furnace efficiency is discussed. The numerical mathematical radiation model seems to be reliable and thus may be extended to investigate combustion-radiation interactions in other industrial applications.

References

[1] Henson J. C., Simulation of submitted in partial spark ignition engines with special emphasis on radiative heat transfer, Ph.D thesis, Loughborough University UK, 1998.
[2] Moghari M., Hosseini S., Shokouhmand H., Sharifi H., Izadpanah S., A numerical study on thermal behavior of a D-type water-cooled steam boiler, Applied Thermal Engineering 37 (2012) 360-372.
[3] Crnomarkovic N., Sijercic M., Belosevic S., Tucakovic D., Zivanovic T., Numerical investigation of processes in the lignite-fired furnace when simple gray gas and weighted sum of gray gases models are used, Int. J. Heat Mass Tran. 56 (2013) 197-205.
[4] Tan C.-K, Jenkis J., Ward J., Broughton J., Heeley A., Zone modelling of the thermal performances of a large-scale bloom reheating furnace, applied thermal engineering 50 (2013) 1111-1118.
[5] Trivic D. N., 3-D radiation modeling of nongray gases–particles mixture by two different numerical methods, Int. J. Heat Mass Tran. 70 (2014) 298-312.
[6] Murty C.V.S., Richter W., Murthy M.V.K., Modeling of thermal radiation in fired heaters, Chem. Eng. Res. Des. 67(1989) 134-144.
[7] Méchi R., Farhat H., Halouani K., Radhouani M.-S., Modélisation des transferts radiatifs dans un incinérateur des émissions polluantes de la pyrolyse du bois, Int. J. Thermal Sci. 43 (2004) 697-708.
[8] Ebrahimi H., Zamaniyan A., Mohammad zadeh J.S.S., Khalili A. A., Zonal modeling of radiative heat transfer in industrial furnaces using simplified model for exchange area calculation, Appl. Math. Modell. 37 (2013) 8004-8015.
[9] Hottel H.C., Sarofim A.F., Radiative Transfer, McGraw-Hill, New York, 1967.
[10] Coelho P.J, Numerical simulation of radiative heat transfer from non-gray gases in three-dimensional enclosures, J. Quantitative Spectrosc. Radiative Transfer 74 (2002) 307-328.

An assessment of some non-gray global radiation models in enclosures

J Meulemans

Saint-Gobain Recherche, 39, quai Lucien Lefranc, B.P. 135, F-93303 Aubervilliers Cedex, France

E-mail: johann.meulemans@saint-gobain.com

Abstract. The accuracy of several non-gray global gas/soot radiation models, namely the Wide-Band Correlated-K (WBCK) model, the Spectral Line Weighted-sum-of-gray-gases model with one optimized gray gas (SLW-1), the (non-gray) Weighted-Sum-of-Gray-Gases (WSGG) model with different sets of coefficients (Smith et al., Soufiani and Djavdan, Taylor and Foster) was assessed on several test cases from the literature. Non-isothermal (or isothermal) participating media containing non-homogeneous (or homogeneous) mixtures of water vapor, carbon dioxide and soot in one-dimensional planar enclosures and multi-dimensional rectangular enclosures were investigated. For all the considered test cases, a benchmark solution (LBL or SNB) was used in order to compute the relative error of each model on the predicted radiative source term and the wall net radiative heat flux.

1. Introduction

Accurate modeling of radiative heat transfer is essential for the improvement of the performance of industrial flame furnaces (thermal efficiency and pollutant emissions). The main thermal radiation issues to be addressed are the radiative properties of combustion products (gases and particles) and their modeling, and the Radiative Transfer Equation (RTE) solution methods. The development of computationally efficient and accurate radiative property models for non-gray gas/soot radiation analysis has been an active research subject for decades (see [1–4] and references therein).

Line-By-Line (LBL) calculations yield exact solutions but millions of RTE resolutions are required. The turnaround time is therefore not acceptable for engineering calculations. Gray gas (GG) models, based on total emittance correlations [5, 6], are computationally efficient and thus widely used for engineering calculations but often yield relatively large errors [7].

Between these two extremes, a vast literature has been devoted to the development of spectral models. The Statistical Narrow Band (SNB) model [8], formulated in average transmissivity or absorption coefficients with a correlated k-distribution method (CK), is the most accurate gas radiative properties model and is considered as a reference model when LBL calculations are not available [9]. Wide Band (WB) models are less accurate than SNB models but more widely used for industrial applications [10]. For combustion applications, global models like the non-gray Weighted-Sum-of-Gray-Gases model (WSGG) [11], the Spectral-Line Weighted-sum-of-gray-gases model (SLW) [12–14] and the Wide Band Correlated-k model (WBCK) [15–17] may yield relatively good predictions in a computationally efficient manner. However, the accuracy assessment highly depends on the radiative transfer problem studied (e.g., type of enclosure, participating species, distributions of temperature and concentrations) [7, 18] and the spectral databases used to compute the LBL solutions [9, 19] and the parameters of the models.

The objective of this work is to investigate the performances of several (non-gray) global radiation models by comparison with benchmark solutions available in the literature. The accuracy of the following models was assessed : the Wide-Band Correlated-K model with nineteen gray gases (WBCK-19) [15–17], the Spectral Line Weighted-sum-of-gray-gases model with one optimized gray gas (SLW-1) [20, 21], the (non-gray) Weighted-Sum-of-Gray-Gases model [11] with coefficients from Smith et al. [22] (WSGG-Smith), Soufiani and Djavdan [23] (WSGG-Soufiani), Taylor and Foster [24] (WSGG-Taylor). A gray gas model (GG), based on total emittance correlations from Leckner [6], was also included in the study. The test problems were selected based on the following criteria : (i) availability of benchmark solutions (LBL or SNB) ; (ii) participating media composed of mixtures of water vapor, carbon dioxide and soot ; (iii) distributions of temperature and concentrations close to those encountered in flame furnaces. To the best of the author's knowledge, the evaluation of the aforementioned global models on the same set of test problems has not been carried out in the past.

The present paper is organized as follows. The second section gives a brief overview of the global models evaluated in this study. The test problems are described in the third section. The obtained results are presented and discussed in the fourth section.

2. Description of global radiation models

2.1. Weighted-Sum-of-Gray-Gases (WSGG) model

The WSGG model [5] expresses the total gas emittance as a weighted sum of gray gas emittances. The emission weighting factors and the absorption coefficients are obtained from a fit to total emittance measurements of isothermal and homogenous columns. The WSGG model is often used in combustion problems assuming that the medium is gray. In such a case, the absorption coefficient of the medium is calculated from the total gas emittance and the mean beam length. The total gas emittance can be computed using the Hottel's charts [5] or the Leckner's correlations [6]. In the present study, we used the Leckner's correlations.

Modest [11] showed that the WSGG approach may be implemented as a non-gray gas model by solving the RTE for N_g gray gases plus one transparent gas. The non-gray WSGG model with the coefficients determined by Smith et al. [22], Soufiani and Dvajdan [23] and Taylor and Foster [24] was used in the present study.

More recent sets of coefficients, based on HITEMP-2010 [25], can be found in [26, 27] while recent methodologies were proposed to derive the coefficients for either different molar ratios [28] or arbitrary concentrations of water vapor, carbon dioxide and soot [29].

2.2. Spectral-Line Weighted-sum-of-gray-gases (SLW) model

The SLW may be regarded as an improved WSGG model, obtained from the line-by-line spectra of the absorbing species [12–14]. For a sufficient number of gray gases, the SLW model yields exact solutions of the gas radiation problem [30, 31]. In the present study, we used the SLW model with one optimized gray gas (SLW-1) [20, 21]. The parameters of SLW-1 were derived from the new correlations for the absorption line blackbody distribution function (ALBDF) of water vapor, carbon dioxide and carbon monoxide developed recently by Pearson et al. [32].

2.3. Wide Band Correlated-k (WBCK) model

WB models yield transmittances and cannot thus be easily be coupled to differential solution methods of the RTE. The WBCK model [15–17] gives an absorption coefficient for each spectral interval. The weight of each spectral interval is calculated from the blackbody fraction of the interval. Since there are several hundreds of wavenumber intervals for a mixture of water vapor and carbon dioxide, a gray gases formulation is used to further decrease the computational effort

Journal of Physics: Conference Series **676** (2016) 012017 doi:10.1088/1742-6596/676/1/012017

by gathering intervals with an absorption coefficient within an interval range. In the present study, the WBCK model with nineteen gray gases was used (WBCK-19).

2.4. Soot radiation model

If soot particles are very small so that the Rayleigh theory applies for all particles and relevant wavelengths, the extinction coefficient is described by :

$$\kappa_\lambda = C_0 \frac{f_v}{\lambda} \tag{1}$$

where C_0 is a constant depending only on the soot index of refraction [3]. The Planck-mean extinction coefficient can be computed as :

$$\kappa_P = 3.83 f_v C_0 T / C_2 \tag{2}$$

where C_2 is the second Planck function constant.

In the present study, the Planck mean absorption coefficient of soot is added to each gray gas for the GG, WBCK and WSGG models using the index of refraction given by Tien and Lee [33]. In the SLW-1 model, soot is treated as a non-gray gas specie for which ALBDF is calculated [20].

3. Description of test problems

The accuracy of WBCK-19, SLW-1, (non-gray) WSGG and GG models was assessed on several test cases from the literature, namely non-isothermal (or isothermal) participating media containing non-homogeneous (or homogeneous) mixtures of water vapor, carbon dioxide and soot in one-dimensional planar enclosures or multi-dimensional rectangular enclosures. All models were not applied to all test cases though (test cases 1 and 2 : GG, WBCK-19, SLW-1 and WSGG-Smith ; test cases 3 and 4 : WSGG-Smith, WSGG-Soufiani and WSGG-Taylor).

Each solution of the GG, WBCK and (non-gray) WSGG models was obtained using the RTE solver (FVM) available in two different commercial CFD codes (GTM-X [34] and ANSYS Fluent [35]). Polar/azimuthal angular quadrature schemes and the step scheme (spatial discretization) are used in both codes. The GG and WBCK models provided in GTM-X were used while the (non-gray) WSGG models were implemented in ANSYS Fluent with user-defined functions.

The exact analytical RTE solution of Solovjov et al. [36] was used for one-dimensional test cases with the SLW-1 model. To the best of the author's knowledge, the SLW-1 cannot yet be readily used in non-isothermal participating media containing non-homogeneous (or homogeneous) mixtures in multi-dimensional enclosures. Issues related to the choice of the reference state and the optimization procedure have yet to be clarified. Therefore, SLW-1 was not implemented in the CFD codes used for this study.

For all the considered test cases, a benchmark solution (LBL or SNB) was used in order to compute the relative error of each model on the predicted radiative source term and the wall net radiative heat flux. Since different computational codes were used in this study, the computational costs associated with each model could not be compared fairly and are thus not reported here.

3.1. Test case 1 : homogeneous and isothermal mixtures in one-dimensional planar enclosures

The first problem (test case 1) was previously studied by Denison [13, 14]. It consists of a homogeneous and isothermal medium containing a mixture of water vapor (40%), carbon dioxide (20%) and soot at 1250 K in one-dimensional planar enclosures. The total gas pressure is 1 atm. The walls are assumed gray ($\epsilon_w = 0.8$) and at 400 K and 1500 K at $x = 0$ and $x = L$, respectively. Four different configurations were investigated by varying the distance between the walls ($L = 0.1,\ 1$ or $3\ m$) and the soot volume fraction ($f_v = 10^{-7}$ or $10^{-6}\ m^3/m^3$).

3.2. Test case 2 : non-homogeneous and non-isothermal mixture in a one-dimensional planar enclosure

The second problem (test case 2) was previously studied by Bressloff [37] and other authors(e.g., [21, 38, 39]). It consists of a non-homogeneous and non-isothermal medium containing a mixture of water vapor, carbon dioxide and soot in a one-dimensional planar enclosure ($L = 1\ m$). The total gas pressure is 1 atm. The walls are assumed black and cold at 0 K. Temperature, gas species (molar fraction) and soot (volume fraction) have the following parabolic distributions :

$$T(x) = 800 + 4000x(L - x) \quad (3)$$

$$X_{CO_2}(x) = 0.06 + 0.4x(L - x) \quad (4)$$

$$X_{H_2O}(x) = 2X_{CO_2} \quad (5)$$

$$f_v(x) = [40x(L - x) + 6] \times f_{v0} \quad (6)$$

where $f_{v_0} = 10^{-7}\ m^3/m^3$ (configuration A, high soot loading) or $10^{-8}\ m^3/m^3$ (configuration B, intermediate soot loading). Figure 1 represents the distributions of the temperature and concentrations fields along the enclosure.

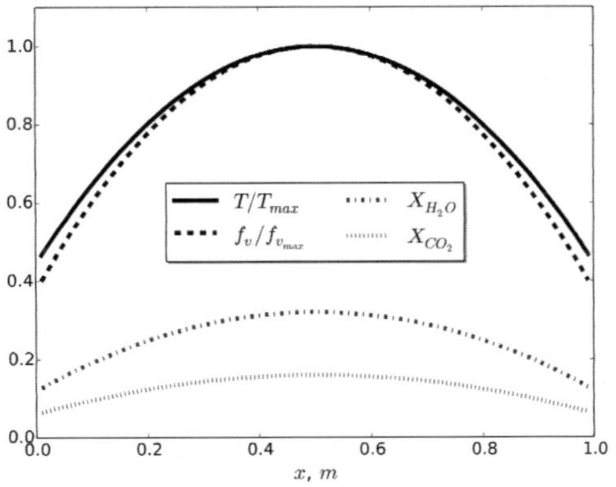

Figure 1. Distributions of the normalized temperature and concentrations fields for test case 2 ($T_{max} = 1800\ K$, $f_{v_{max}} = 16\ f_{v0}$).

3.3. Test case 3 : homogeneous and non-isothermal mixture in a two-dimensional rectangular enclosure

The third problem (test case 3) was previously studied by Goutière et al. [18] and other authors (e.g., [15, 40]). It consists of a homogeneous and non-isothermal medium containing a mixture of water vapor (20%) and carbon dioxide (10%) in a two-dimensional rectangular enclosure ($1 \times 0.5\ m^2$). The total gas pressure is 1 atm. The walls are assumed black and cold at 0 K. The temperature distribution is defined as follows :

$$T(x, y) = \begin{cases} 800 + (14000x - 400)\left(1 - 3y_0^2 + 2y_0^3\right), & \text{if } x \leq 0.1\ m \\ 800 - \frac{10000}{9}(x - 1)\left(1 - 3y_0^2 + 2y_0^3\right), & \text{if } x \geq 0.1\ m \end{cases} \quad (7)$$

with $y_0 = |0.25 - y|/0.25$. The temperature reaches a maximum (1800 K) along the centerline at $x = 0.1$ m. Figure 2 represents the distribution of the temperature field in the rectangular enclosure.

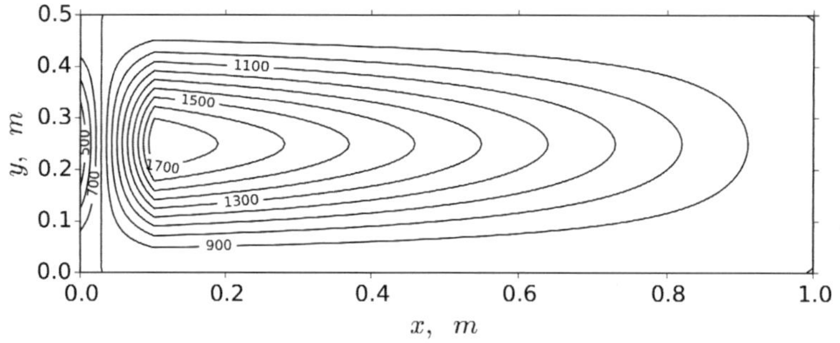

Figure 2. Distribution of the temperature field (in K) for test case 3.

3.4. Test case 4 : homogeneous and non-isothermal mixture in a three-dimensional rectangular enclosure
The fourth problem (test case 4) was previously studied by Liu [41] and other authors (e.g., [42–44]). It consists of a homogeneous and non-isothermal medium containing a mixture of water vapor (20%) and carbon dioxide (10%) in a three-dimensional rectangular enclosure (2×2×4 m^3). The total gas pressure is 1 atm. The walls are assumed black and cold at 300 K. The temperature is non-uniform but symmetrical along the centerline of the enclosure :

$$T(r) = (T_c - T_e)\, f(r/R) + T_e \quad (8)$$

$$f(r/R) = 1 - 3\,(r/R)^2 + 2\,(r/R)^3 \quad (9)$$

where T_c, T_e, r and R are the temperature along the centerline of the enclosure, the exit temperature at $z = 4$ m, the distance from the enclosure centerline and the radius of the circular region ($R = 1$ m), respectively. The temperature outside the circular region is assumed to be uniform and at the value of the exit temperature. The centerline temperature is assumed to increase linearly from 400 K at the inlet ($z = 0$ m) to 1800 K at $z = 0.375$ m, then decreases linearly to 800 K at the exit. Figure 3 represents the distribution of the temperature field in the rectangular enclosure.

4. Results and discussion

4.1. Test case 1 : homogeneous and isothermal mixtures in one-dimensional planar enclosures
LBL calculations from Denison [13, 14] were used as benchmark solutions. Figure 4 shows the predicted profiles of the radiative source term. The S-shaped profiles obtained with the benchmark solutions (LBL) were well reproduced by all models but the gray gas model. The cold wall net radiative heat flux and the relative error by comparison with LBL benchmark data are reported in Table 1. The results obtained with the models were almost exact (relative errors less than 3%) with the noticeable exception of the gray gas model (relative errors greater than 7.5% for all configurations but one).

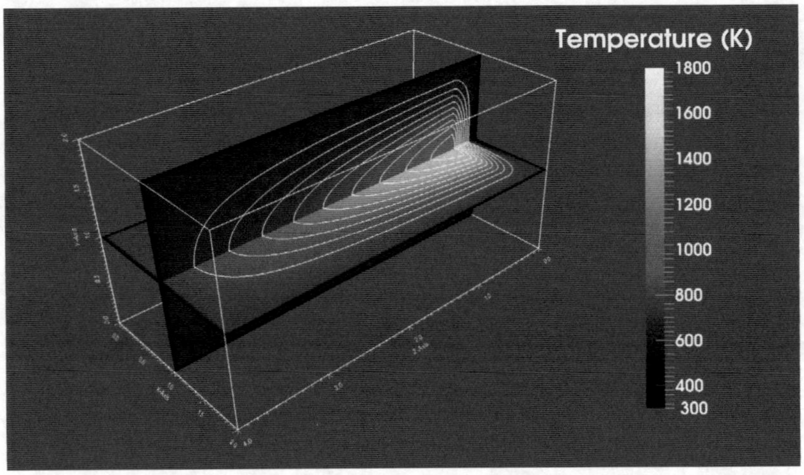

Figure 3. Distribution of the temperature field for test case 4.

Table 1. Results of the cold wall net radiative heat flux for test case 1 : $L = 0.1\ m$, no soot (configuration A) ; $L = 3\ m$, no soot (configuration B) ; $L = 0.1\ m$, $f_v = 10^{-7}\ m^3/m^3$ (configuration C) and $L = 1\ m, f_v = 10^{-6}\ m^3/m^3$ (configuration D). Relative errors greater than 5% are shaded.

Model	Config. A		Config. B		Config. C		Config. D	
	$q\ (kW/m^2)$		$q\ (kW/m^2)$		$q\ (kW/m^2)$		$q\ (kW/m^2)$	
LBL (benchmark) [13, 14]	184.0		162.0		181.0		112.0	
WBCK-19	183.5	-0.3%	160.3	-1.0%	181.4	0.2%	115.1	2.8%
SLW-1	183.2	-0.4%	157.5	-2.8%	180.3	-0.4%	113.8	1.6%
WSGG-Smith	181.4	-1.4%	163.5	0.9%	179.2	-1.0%	114.0	1.8%
Leckner	169.1	-8.1%	135.6	-16.3%	167.5	-7.5%	113.7	1.5%

4.2. Test case 2 : non-homogeneous and non-isothermal mixture in a one-dimensional planar enclosure

SNB calculations from Bressloff [37] were used as benchmark solutions. The results obtained by Bressloff with the (non-gray) WSGG model using the coefficients of Truelove [45] are also given. Figure 5 shows the predicted profiles of the radiative source term. The M-shaped profiles obtained with the benchmark solutions (SNB) were well reproduced by all models. The maximum discrepancy occurred at the center and at the vicinity of the cold walls. The relative errors by comparison with SNB benchmark data are reported in Table 2. The relative error was computed as follows :

$$\delta = |\nabla \cdot \mathbf{q}_{\text{SNB}} - \nabla \cdot \mathbf{q}| \,/\, |\nabla \cdot \mathbf{q}_{\text{SNB}}|_{\max} \tag{10}$$

WSGG-Truelove yielded the most accurate results among the models tested (mean and maximum relative errors less than 3% and 5%, respectively). The results obtained with SLW-1 and GG (Leckner) were in very good agreement with the benchmark solution only for configuration B (mean and maximum relative errors less than 3% and 5%, respectively). The relatively good results obtained with the gray gas model on configuration B were not expected

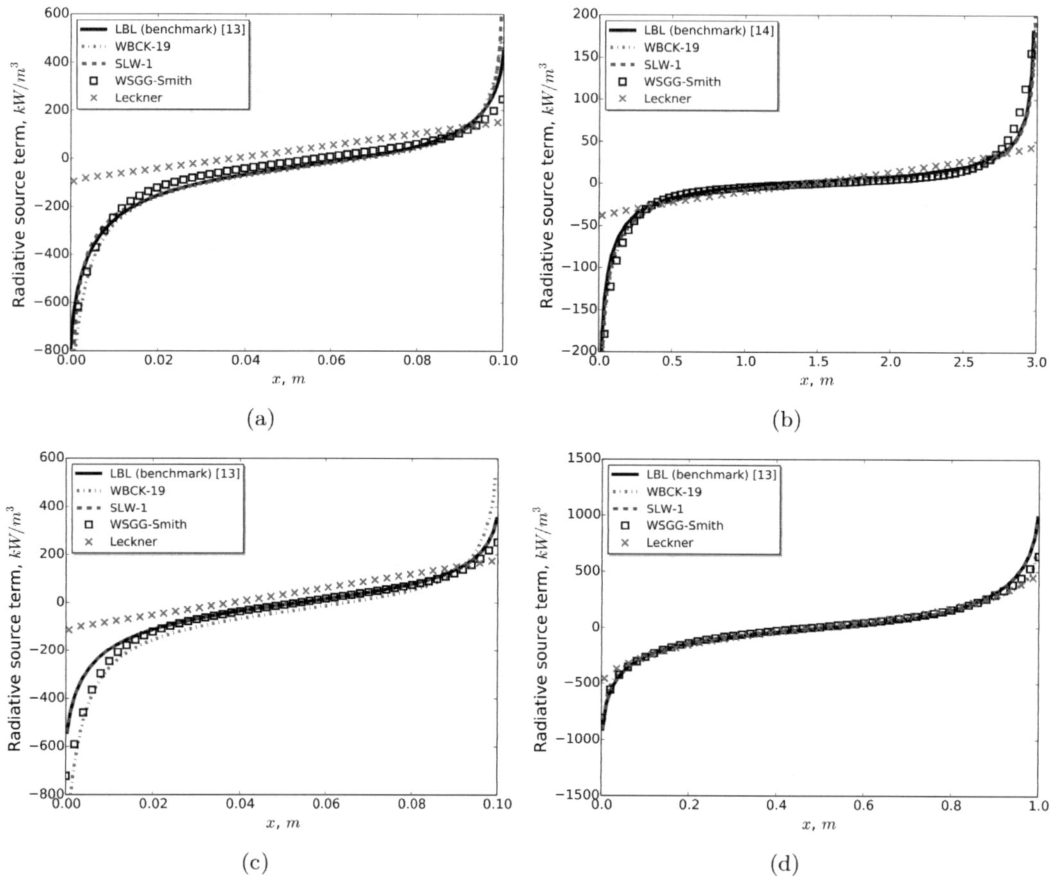

Figure 4. Predicted profiles of radiative source term compared with the LBL benchmark data [13, 14] for test case 1 : (a) $L = 0.1$ m, no soot (configuration A) ; (b) $L = 3$ m, no soot (configuration B) ; (c) $L = 0.1$ m, $f_v = 10^{-7}$ m^3/m^3 (configuration C) and (d) $L = 1$ m, $f_v = 10^{-6}$ m^3/m^3 (configuration D).

and might be a coincidence. For all the models tested, the non-gray behavior of soot seemed to be the main source of error (larger discrepancies with the benchmark solution were found on configuration A vs. configuration B). This aspect should be further investigated.

4.3. Test case 3 : homogeneous and non-isothermal mixture in a two-dimensional rectangular enclosure

LBL calculations from Chu et al. [40] were used as benchmark solutions. The results obtained by Chu et al. [40] with SNB-CK and Goutière et al. [18] with SNB are also reported. The results presented below were computed on a cartesian uniform mesh composed of 200 $\times$ 100 grid cells and 144 discrete ordinates ($N_\theta = N_\phi = 6$). The influence of the spatial and angular discretization on the results (radiative source term and wall radiative heat flux) was thoroughly verified [46]. Figure 6 shows the predicted wall net radiative heat flux distributions on the top and right walls of the enclosure calculated using different numbers of discrete ordinates with WSGG-Smith. The location of the maximum on the top wall slightly shifted to the right of the enclosure with finer angular discretizations. The shape of the predicted profile on the right wall was very sensitive to the angular discretization used. With a coarse angular discretization (i.e.,

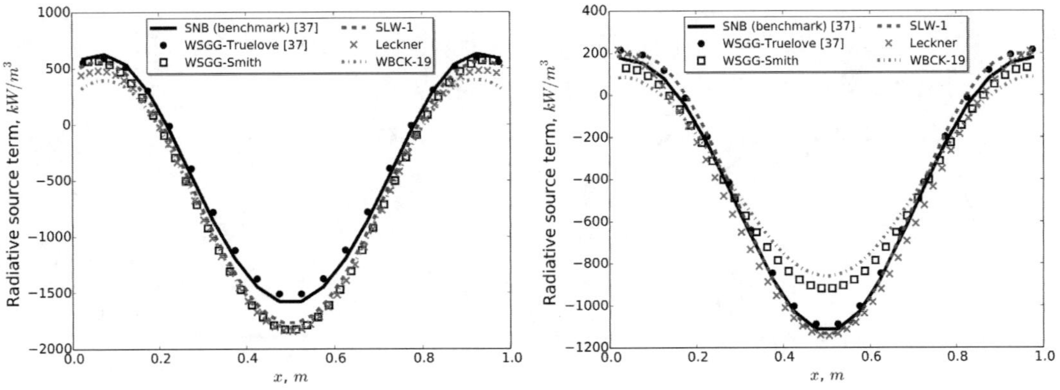

(a) (b)

Figure 5. Predicted profiles of radiative source term compared with the SNB benchmark data [37] for test case 2 : (a) $f_{v0} = 10^{-7}\ m^3/m^3$ (configuration A) and (b) $f_{v0} = 10^{-8}\ m^3/m^3$ (configuration B).

Table 2. Relative errors by comparison with SNB benchmark data [37] for test case 2 : $f_{v0} = 10^{-7}\ m^3/m^3$ (configuration A) and $f_{v0} = 10^{-8}\ m^3/m^3$ (configuration B). Mean and maximum relative errors greater than 5% and 10% respectively are shaded.

Model	Config. A		Config. B	
	δ (%)		δ (%)	
	mean	max.	mean	max.
WSGG-Truelove [37]	3.2	4.9	2.8	3.6
WBCK-19	11.6	15.9	10.6	23.4
SLW-1	6.5	11.3	2.6	5.1
WSGG-Smith	10.0	14.9	7.4	18.0
Leckner	11.0	15.6	4.6	8.0

less than 64 discrete ordinates, $N_\theta = N_\phi = 4$), the profile was M-shaped whereas the converged profile obtained with a fine angular discretization ($N_\theta = N_\phi = 6$) was U-shaped and rather flat at the center. Similar results were obtained with WSGG-Soufiani and WSGG-Taylor but are not reported here for the sake of brevity. Besides, using a finer mesh composed of 400 × 200 grid cells did not alter the results (not reported here).

Figure 7 shows the predicted radiative source term along the enclosure centerline in the x and y directions. The profiles obtained with the benchmark solutions (LBL) were well reproduced by the WSGG model. The location of the minimum value of the radiative source term along the centerline in the x direction was exact and corresponded to the maximum of emission (maximum of gas temperature). The M-shaped profile along the centerline in the y direction was rather well predicted by the models as well. The maximum discrepancy occurred at the center. Relative errors on the radiative source term along the enclosure centerline in the x direction and y direction are reported in Table 3. The main difference between the models lied in the prediction of the minimum of the radiative source term along the centerline in the x direction. While WSGG-Smith was close to the benchmark data (relative error less than 6%),

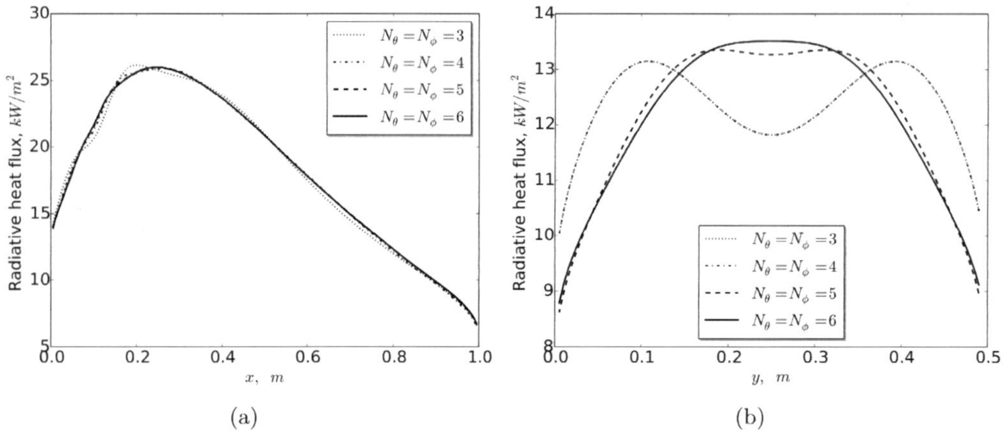

Figure 6. Comparison of wall net radiative heat flux distributions on the top, (a), and right, (b), walls of the enclosure calculated using different angular discretizations for test case 3.

both WSGG-Soufiani and WSGG-Taylor exhibited discrepancies greater than 20%.

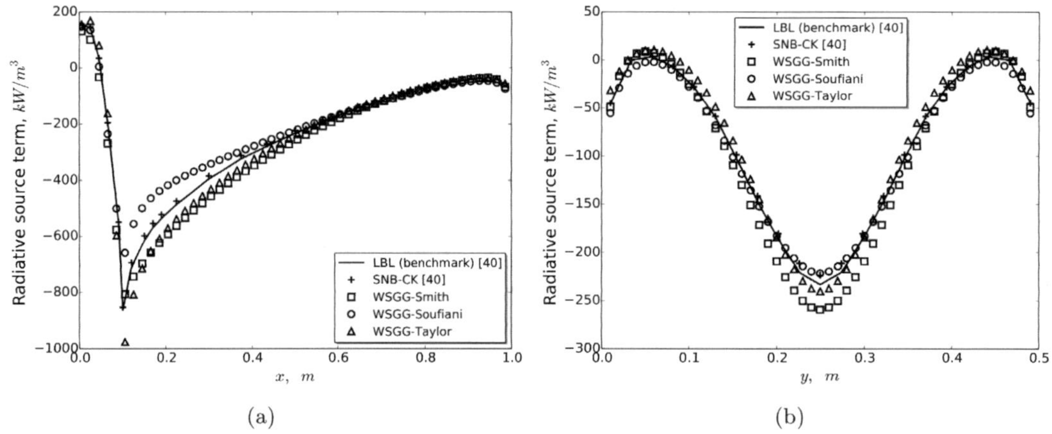

Figure 7. Predicted profiles of radiative source term along the enclosure centerline in the x direction, (a), and y direction, (b), compared with the LBL benchmark data [40] for test case 3.

Results of the net wall radiative heat fluxes at the center of the top and right walls and the radiative source term at the center of the enclosure are reported in Table 4. Based on these local values, WSGG-Soufiani appeared to yield the most accurate results among the models tested (relative errors less than 5%). The SNB results of Goutière et al. are close to the benchmark data except for the predicted net radiative heat flux at the center of the top wall (relative error greater than 20%). This unexpected discrepancy might be due to differences in the spatial/angular discretizations used by the authors.

Table 3. Relative errors on the radiative source term along the centerline by comparison with the LBL benchmark data [40] for test case 3. Mean and maximum relative errors greater than 5% and 10% respectively are shaded.

Model	δ_x (%)		δ_y (%)	
	mean	max.	mean	max.
SNB-CK [40]	0.8	1.9	1.4	4.1
WSGG-Smith	5.7	12.8	5.9	14.9
WSGG-Soufiani	6.0	22.3	3.4	7.0
WSGG-Taylor	5.6	21.5	3.7	8.8

Table 4. Results of the net wall radiative heat fluxes at the center of the right and top walls and the radiative source term at the center of the enclosure compared with the LBL benchmark data [40] for test case 3. Relative errors greater than 10% are shaded.

Model	q (kW/m^2)		q (kW/m^2)		$-\nabla \cdot \mathbf{q}$ (kW/m^3)	
	At (1 m, 0.25 m)		At (0.5 m, 0.5 m)		At (0.5 m, 0.25 m)	
LBL (benchmark) [40]	13.053		17.881		-233.090	
SNB-CK [40]	12.572	-3.7%	17.222	-3.7%	-223.290	-4.2%
SNB [18]	12.668	-2.9%	21.630	-21.0%	-226.000	-3.0%
WSGG-Smith	13.509	3.5%	20.821	16.4%	-259.416	11.3%
WSGG-Soufiani	13.636	4.5%	18.572	3.9%	-221.797	-4.8%
WSGG-Taylor	11.834	-9.3%	17.445	-2.4%	-240.333	3.1%

4.4. Test case 4 : homogeneous and non-isothermal mixture in a three-dimensional rectangular enclosure

SNB calculations from Liu [41] were used as benchmark solutions. The results presented below were computed on a cartesian uniform mesh composed of 40 × 40 × 80 grid cells and 72 discrete ordinates ($N_\theta = N_\phi = 3$). The influence of the spatial and angular discretization on the results (radiative source term and wall radiative heat flux) was thoroughly verified. Grid independence of the results is demonstrated in Figure 8 by comparing the results of the radiative source term along the centerline (1 m, 1 m, z) and the wall net radiative heat flux along (2 m, 1 m, z) calculated using three different grids with WSGG-Smith. The effect of grid size on the results was found negligible except at the location of the minimum of the radiative source term and the maximum of the wall net radiative heat flux. Unlike test case 3, no influence of the angular discretization was found (not reported here). Similar results were obtained with WSGG-Soufiani and WSGG-Taylor but are not reported here for the sake of brevity.

Figure 9 shows the predicted of radiative source term and wall radiative heat flux. The profiles obtained with the benchmark solutions (SNB) were qualitatively reproduced by the (non-gray) WSGG model. The minimum of the radiative source term along the centerline was well captured although its intensity varied notably with the coefficients used for the predictions. The location of the peak of the net radiative heat flux was not well predicted while its magnitude depended on the coefficients used.

Relative errors by comparison with the SNB benchmark data are reported in Table 5. The relative errors were computed as follows :

$$\delta_\phi = |\Phi_{\mathrm{SNB}} - \Phi| \,/\, |\Phi_{\mathrm{SNB}}|_{\max} \tag{11}$$

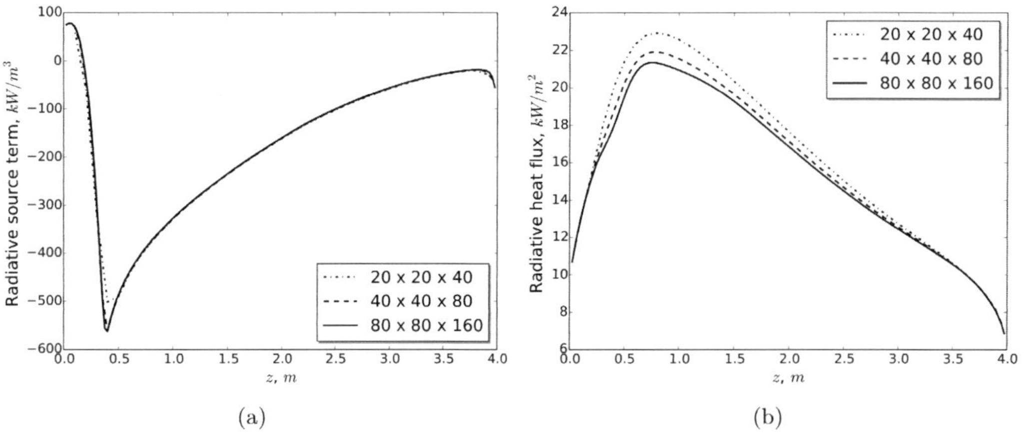

Figure 8. Comparison of radiative source term and wall radiative heat flux distributions of the enclosure calculated using three computational grids with WSGG-Smith for test case 4 : (a) radiative source term along the centerline (1 m, 1 m, z) ; wall radiative heat flux along (2 m, 1 m, z).

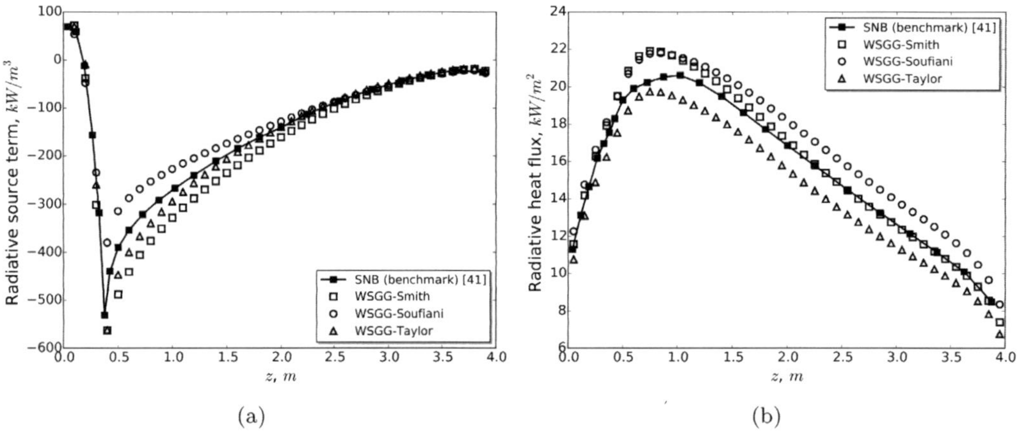

Figure 9. Predicted profiles of radiative source term and wall radiative heat flux compared with the SNB benchmark data [41] for test case 4 : (a) radiative source term along the centerline (1 m, 1 m, z) and (b) wall radiative heat flux along (2 m, 1 m, z).

where $\Phi = \nabla \cdot \mathbf{q}$ or q. The mean relative errors were less than 7.5% for all the models tested. The maximum relative errors were between 17% and 30%, indicating that the (non-gray) WSGG model failed to yield accurate local predictions, mostly at the location of the peak of gas temperature.

5. Conclusion

Three non-gray global gas radiation models were employed to solve radiative transfer problems in enclosures. For all the considered test cases, a benchmark solution (LBL or SNB) was used in order to compute the relative error of each model on the predicted radiative source term and the wall net radiative heat flux.

Table 5. Relative errors by comparison with the SNB benchmark data [41] for test case 4. Mean and maximum relative errors greater than 5% and 10% respectively are shaded.

Model	δ_q (%)		$\delta_{\nabla\cdot\mathbf{q}}$ (%)	
	mean	max.	mean	max.
WSGG-Smith	2.7	8.0	7.2	20.3
WSGG-Soufiani	5.4	7.2	6.1	29.1
WSGG-Taylor	5.2	6.5	3.8	17.2

The following conclusions were drawn from the analysis carried out :

1. SLW-1 and WSGG yielded reasonable overall accuracy in one-dimensional planar enclosures. WBCK-19 did not prove to be more accurate than SLW-1 and WSGG on the test problems investigated.
2. WSGG failed to accurately predict local quantities in multi-dimensional enclosures. The optimal set of coefficients depended on the test case investigated and the quantity looked at to compute the relative error with the benchmark data. This was clearly not satisfactory.
3. The non-gray modeling of soot should be further investigated (e.g., the influence of the optical constant of soot on the predictions).
4. Spatial and angular independent solutions were computationally costly with the FVM implemented in the two CFD codes used for this study. This might preclude the use of WSGG-like models with a number of gray gases greater than three or four for practical industrial applications.

Further work may include : (i) additional test cases, namely mixtures with different (constant or variable) molar ratios (e.g., [13]), multi-dimensional enclosures containing soot (e.g., [39]) or non-homogeneous mixtures ; (ii) accuracy assessment of the updated coefficients for WSGG and (iii) investigation of the combined use of global models and P-1 in multi-dimensional enclosures.

Acknowledgments

Dr. Luuk Thielen (CelSian Glass & Solar) prepared the special executables of the CFD code used to test the WBCK model. Professor Vladimir Solovjov (Brigham Young University) provided the tabulated data of Bressloff (SNB and WSGG-Truelove) and Denison (LBL and SLW). Professor Solovjov also provided the SLW-1 model (Maple codes).

References

[1] Taine J and Soufiani A 1999 *Adv. Heat Transfer* **33** 295–414
[2] Howell J R, Siegel R and Mengüç M P 2010 *Thermal Radiation Heat Transfer* 5th ed (Boca Raton, FL: CRC Press, Inc.)
[3] Modest M 2013 *Radiative Heat Transfer* 3rd ed (New York: Academic Press)
[4] Modest M 2013 *J. Heat Transfer* **135** 061801–1–12
[5] Hottel H and Sarofim A 1967 *Radiative Transfer* (New York: McGraw-Hill)
[6] Leckner B 1972 *Combust. Flame* **19** 33–48
[7] Goutière V, Charette A and Kiss L 2002 *Numer. Heat Transfer, Part B* **41** 361–81
[8] Soufiani A and Taine J 1997 *Int. J. Heat Mass Transfer* **40** 987–91
[9] Chu H, Liu F and Zhou H 2011 *Int. J. Heat Mass Transfer* **54** 4736–45
[10] Edwards D 1976 *Advances in Heat Transfer* **12** 115–93
[11] Modest M 1991 *J. Heat Transfer* **113** 650–56
[12] Denison M and Webb B 1993 *J. Heat Transfer* **115** 1004–11

[13] Denison M 1994 *A Spectral Line-based Weighted-sum-of-gray-gases model for arbitrary RTE solvers* Ph.D. thesis Brigham Young University
[14] Denison M and Webb B 1995 *J. Heat Transfer* **117** 359–65
[15] Ströhle J and Coelho P 2002 *Int. J. Heat Mass Transfer* **45** 2129–39
[16] Ströhle, J, Schnell, U and Hein, KRG 2003 Optimisation of correlated-k method for non-gray radiative transfer calculations in multidimensional enclosures *Proc. of Eurotherm 73 on Computational Thermal Radiation in Participating Media, April 15-17, 2003, Mons, Belgium*
[17] Ströhle J 2008 *J. Quant. Spectr. Rad. Transfer* **109** 1622–40
[18] Goutière V, Liu F and Charette A 2000 *J. Quant. Spectrosc. Radiat. Transfer* **64** 299–326
[19] Poitou D and André F 2013 *Int. J. Therm. Sci.* **64** 11–21
[20] Solovjov V, Lemonnier D and Webb B 2011 *J. Quant. Spectrosc. Radiat. Transfer* **112** 1205–12
[21] Solovjov V, Lemonnier D and Webb B 2011 *J. Heat Transfer* **133** 102701–1–9
[22] Smith T, Shen Z and Friedman J 1982 *J. Heat Transfer* **104** 602–8
[23] Soufiani A and Djavdan E 1994 *Combust. Flame* **97** 240–50
[24] Taylor P and Foster P 1975 *Int. J. Heat Mass Transfer* **18** 1331–32
[25] Rothman L, Gordon I, Barber R, Dothe H, Gamache R, Goldman A, Perevalov V, Tashkun S and Tennyson J 2010 *J. Quant. Spectr. Rad. Transfer* **111** 2139–50
[26] Kangwanpongpan T, França F H, Corrêa da Silva R, Schneider P S and Krautz H J 2012 *Int. J. Heat Mass Transfer* **55** 7419–33
[27] Dorigon L J, Duciak G, Brittes R, Cassol F, Galarça M and França F H 2013 *Int. J. Heat Mass Transfer* **64** 863–73
[28] Johansson R, Leckner B, Andersson K and Johnsson F 2011 *Combust. Flame* **158** 893–901
[29] Cassol F, Brittes R, França F H and Ezekoye O A 2014 *Int. J. Heat Mass Transfer* **79** 796–806
[30] Solovjov V P and Webb B W 2011 *J. Heat Transfer* **133** 042701–1–9
[31] Solovjov V P, Lemonnier D and Webb B W 2014 *J. Quant. Spectrosc. Radiat. Transfer* **143** 83–91
[32] Pearson J T, Webb B W, Solovjov V P and Ma J 2014 *J. Quant. Spectrosc. Radiat. Transfer* **138** 82–96
[33] Lee S and Tien C 1981 Optical constants of soot in hydrocarbon flames *Symp. (Int.) on Combustion* vol 18 pp 1159–66
[34] CelSian 2012 *GTM-X User Manual Version v4.4.6* CelSian Glass & Solar BV
[35] Ansys 2009 *ANSYS FLUENT 12.0 Theory Guide* Ansys, Inc.
[36] Solovjov V P and Webb B W 2008 *J. Quant. Spectrosc. Radiat. Transfer* **109** 245–57
[37] Bressloff N 1999 *Int. J. Heat Mass Transfer* **42** 3469–80
[38] Solovjov V and Webb B 2001 *J. Heat Transfer* **123** 450–57
[39] Demarco R, Consalvi J, Fuentes A and Melis S 2011 *Int. J. Therm. Sci.* **50** 1672–84
[40] Chu H, Liu F and Zhou H 2012 *Int. J. Therm. Sci.* **59** 66–74
[41] Liu F 1999 *J. Heat Transfer* **121** 200–3
[42] Coelho P 2002 *J. Quant. Spectr. Rad. Transfer* **74** 307–28
[43] Trivic D N 2004 *Int. J. Heat Mass Transfer* **47** 1367–82
[44] Selçuk N and Doner N 2009 *J. Quant. Spectr. Rad. Transfer* **110** 184–91
[45] Truelove J 1975 *The Zone Method for Radiative Heat Transfer Calculations* AERE-R (Heat Transfer and Fluid Flow Service, AERE)
[46] Roache P 2009 *Fundamentals of Verification and Validation* (Socorro, NM: Hermosa Publishers)

Study of a V-shape flame based on IR spectroscopy and IR imaging

J. Pouplin, A. Collin, Z. Acem, G. Parent, P. Boulet [1]
LEMTA, CNRS, Université de Lorraine
2 av. de la Forêt de Haye, TSA 60604 – 54518 Vandœuvre-lès-Nancy Cedex, France
E-mail: pascal.boulet@univ-lorraine.fr

P. Vena, C. Galizzi, M. Kühni, F. André, D. Escudié
CETHIL, Campus LyonTech La Doua, INSA de Lyon – CNRS – UCBL - Bât. Sadi Carnot
9, rue de la Physique – 69621 Villeurbanne Cedex

Abstract. Applicability of an IR imaging/spectroscopy diagnostic was tested on a laboratory-scale flame. For this purpose, measurements were carried out on a V-shape flame developed along a wall, with the aim of evaluating the wall temperature and of identifying the flame properties (temperature and species concentrations). Infrared measurements with a multiband camera and a spectrometer were post-processed and compared, in particular, with thermocouple measurements carried out for the wall temperature. Simple evaluation involving a correction for the emissivity showed a quite good agreement when assessed against experimental data. An attempt to reconstruct a flame emission spectrum was also carried out, expecting a possible inverse identification of the flame properties. The method showed a promising behaviour on synthetic data built with a radiative transfer model for gas and wall radiation. However, the spectrum reconstruction method is not yet accurate enough to allow an identification of the flame properties in full confidence when applied to actual experimental data. First tests showed a correct qualitative behaviour, but model refinements are required at least for the flame radiation, before getting accurate flame properties.

1. Introduction

This study was carried out to test the applicability of an IR imaging/spectroscopy diagnostic to laboratory-scale flames. The diagnostic system was developed by LEMTA in Nancy, and experiments were conducted at the CETHIL in Lyon on a premixed air/methane V-flame. The CETHIL has a wide experience of the combustion in a V-shape flame. This is one of the most common experimental configuration for studying premixed combustion in academic conditions. A premixed flow air / methane is produced in a vertical wind tunnel. The flame is anchored behind a rod and expands freely in the output of the device without confinement [1-2]. Wall-flame interactions were the primary focus of the present study which was mainly aimed at evaluating the surface temperature on the wall where the flame was developing. Such measurement with thermocouples is difficult because of the complex combined transfer at the surface. The CETHIL developed very fine micro-thermocouples, with low inertia, but a confirmation was sought on the provided values. Such verification can be obtained

[1] To whom any correspondence should be addressed.

through infrared measurement. The wall temperature is especially uneasy to evaluate because of the combined contributions of radiation from the flame and from the wall itself, convection at the wall surface and conduction inside the wall thickness. Reflection from the background may be also included as the wall temperature is not expected to reach high values since it is cooled by the inlet air-methane mixing. For this purpose, infrared measurements were first analysed in chosen spectral ranges in order to avoid as much as possible any perturbation by the flame. Then, a model was developed and numerical computations were carried out in order to reconstruct all the radiation measured by a FTIR spectrometer along a line of sight, including the background radiation from the wall itself and the complex radiation from the flame. Such attempts to identify thermal radiative properties based on spectrum reconstruction was already presented by different authors [3-5]. The present study follows the same idea. Experiments were conducted on the setup developed by the CETHIL, involving flame shape reconstruction thanks to laser tomography, infrared flux measurements and data obtained from micro-thermocouples inserted in the wall. These measurements provide a good description of the various thicknesses involved in the numerical simulation and also an idea of the temperature range for the wall, avoiding a blind prediction. Actually, the present flame is quite complicated due to its particular shape, but the combustion conditions are well controlled, which makes the present exercise an interesting test for our method of infrared characterization.

2. Experimental setup and methodology

Lean, premixed methane/air V-flames were stabilized at the exit of a 115 x 115 mm square exit nozzle burner. Two 200 × 260 mm ceramic plates were superimposed vertically and oriented parallel to the main flow direction (x-axis in Figure 1). The equivalence ratio of the reactants was 0.6, and the mean flow velocity along the exit plane of the burner was 5 m/sec. One laminar and two turbulent flow settings (turbulent intensity 4% and 8%) were studied. Additional details of the apparatus are provided in [6]. Flame topology data were obtained from Mie scattering of olive oil droplets using a Photron FASCAM SA4 1024 × 1024 pixel high speed CCD and CNI MGL-N-532 continuous wave (CW) laser. The oil droplets were generated with a PIVTEC atomizer. Images had a projected spatial resolution of 5 pixels/mm and represented a 200 x 200 mm region of interest in the flame. 500 instantaneous images were acquired for each flame condition and used to determine the mean position of the reactants, highlighted in Figure 2 as the black iso-contour. This was subsequently used, to subdivide the flow "domain " into reactants, products, and free stream air and correct the separately acquired spectrometer data for variations in mixture composition along the instrument's line-of-sight. Radiative fluxes were measured with a BRUKER FTIR spectrometer and an ORION infrared camera FLIR SC7000 used with a filter at 3,9μm (a specific wavelength such that no emission occurs by combustion gases, allowing to observe the wall behind the flame throughout the flame). Details on this setup may be found in [7-8]. A typical IR image is shown in Figure 3. Data are given in terms of blackbody equivalent temperature. The wall temperature is increasing with vertical position, as the flame get closer to the wall. Five small white circles are superimposed on the image corresponding to five thermocouple locations, referred to T1 to T5 (from the lowest one at axial position x = 60 mm above the leading edge of the plate to the highest one at axial position x = 220 mm). Only four of these vertical positions were considered for the line of sight used by the spectrometer, as the measurement could not be done on the lowest position. The spectrometer was located in front of the wall, observing a typical area shown by the large white circle in Figure 3 (centred around T5 position in the IR image of Figure 3). The typical shape of the registered spectra is shown in Figure 4 (after post-processing considering a reference spectrum obtained with a high temperature blackbody). Actually three spectra are provided in this case of acquisition in front of the T2 position with a flame obtained for a 4% turbulent premixing of air and methane. They are plotted as rough acquisition data in terms of equivalent blackbody intensities (only converted in intensity with no correction for the wall emissivity and reflectivity effect). As can be seen, the repeatability is quite good and spectra seem to be perfectly superimposed. Peaks due to CO_2 and H_2O are obvious, as expected in such spectra where the

Journal of Physics: Conference Series **676** (2016) 012018 doi:10.1088/1742-6596/676/1/012018

combustion products are emitting radiation. There is no strong continuous background in the spectrum, except in the low wavenumber range. This is due to the fact that no soot are produced in the flame. Actually, the signal in the low wavenumber range is mainly due to the wall emission, showing that the evaluation of the wall temperature seems to be possible. A comparison with a blackbody spectrum at 420 K is shown in the Figure in order to give an idea of the emission level (which is confirmed by the micro-thermocouples).

Wall emissivity is a crucial data for an evaluation of the real temperature. It was measured on a separate setup with a dedicated spectrometer (Vertex 80 by BRUKER) and an integrated sphere. The spectrum was analyzed, showing a relatively weak spectral evolution as the wall is actually made of ceramic with a black paint on the surface. Planck's average values between 0.94 and 0.95 were found for blackbody sources with temperature in the range 400-500 K (the relevant range for the wall temperature). A particular value of 0.89 was obtained for the wavelength 3.9 μm (filter of the camera) and the same average value was found for wavenumbers between 2500 and 2700 cm^{-1} (which is in the same range [3.7 - 4.0 μm] and which was of particular interest for some data processing as explained below).

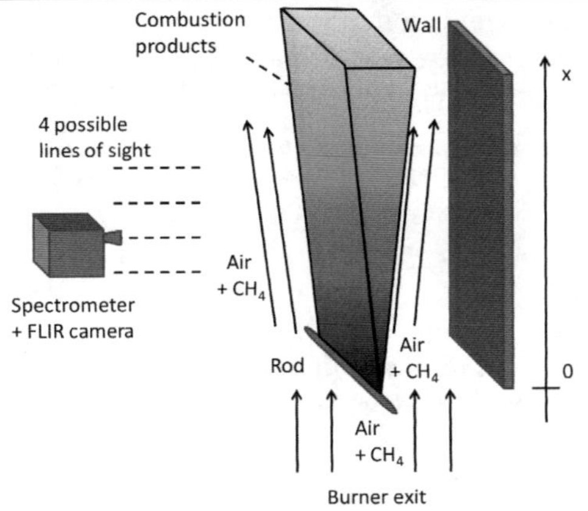

Figure 1: sketch of the setup

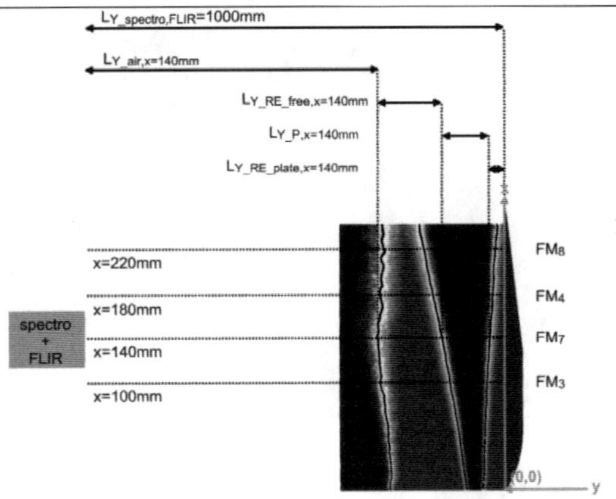

Figure2: visualization of the V shape flame from laser tomography analysis. Oil droplet concentration in false colors (maximum concentration in red, zero concentration in blue)

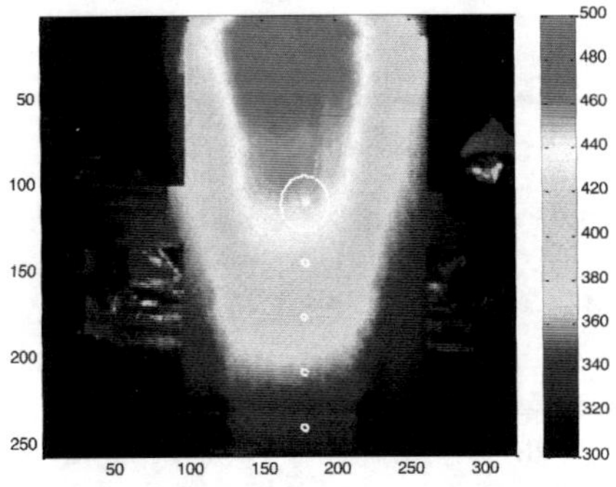

Figure 3: Typical IR image of the wall observed throughout the flame. Observation at 3.9µm. Data converted in equivalent blackbody temperature (in K on the color bar).

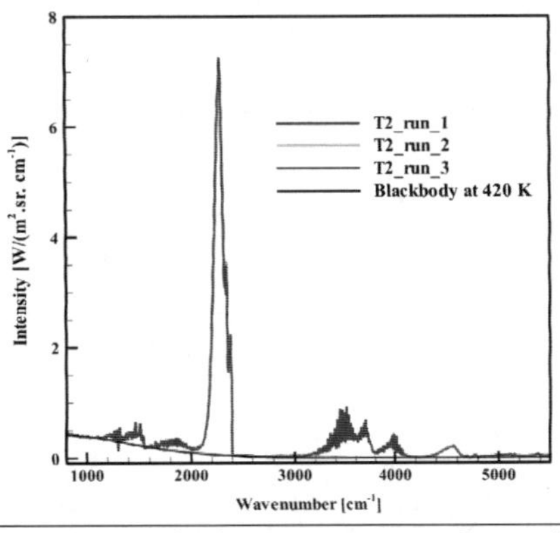

Figure 4: Typical IR spectrum. Intensities are plotted for three measurement tests. Case of turbulence 4%.

3. Simple identification of the wall temperature

Camera images were post-processed in a quite standard manner involving a correction of the temperature with the emissivity 0.89 in order to account for the deviation from perfect emission by a blackbody and also to bring a correction of the signal due to possible reflection of radiation by the wall from the surroundings. Infrared spectra were also used for a similar first evaluation of the wall temperature, using the experimental data obtained with a resolution of 4 cm^{-1}. The part of the spectrum between 2500 and 2700 cm^{-1} was chosen for a fit of the intensity considering the following simple model:

$$I = \epsilon_v I_{0v}(T_{wall}) + (1 - \epsilon_v) I_{0v}(T_{background}) \quad (1)$$

where $I_{0v}(T)$ stands for the spectral blackbody intensity at temperature T. The subscript *background* refers to the surrounding temperature which was set as the ambient one (close to 20°C). This spectral range was chosen to avoid any perturbation by the flame (no soot in the flame and no gas influence in this spectral interval). Then, a least mean square method was used to identify the best wall temperature yielding the right intensity profile in this spectral range.

Table 1 gives a comparison between wall temperatures evaluated from spectroscopic measurements, camera image processing and thermocouples. The temperature range 400 to 550 K is confirmed by all the measurement methods. The influence of the premixing conditions is out of the scope of the present paper, but it is also interesting to see that the complex evolution of the temperature is well captured by all the methods (decrease of the temperature level for low turbulence intensity as compared to the laminar case and re-increase for higher turbulence level). Although some discrepancies may appear between the methods for the laminar case, the agreement is quite correct in the two turbulent cases, especially for such a first blind test, that could be now refined.

Table 1. Wall surface temperature (in K) at different positions according to camera images, to spectrum fit between 2500 and 2750 cm^{-1} and to micro-thermocouples, for 4 positions on the wall (distance from the leading edge of the plate indicated in mm) and 3 flow conditions of the premixing of air and methane.

	laminar			turbulent 4%			turbulent 8%		
	Camera	Spectro	TC	Camera	Spectro	TC	Camera	Spectro	TC
T2(100mm)	480	478	444	425	425	418	479	464	463
T3(140mm)	514	503	475	450	449	447	518	499	500
T4(180mm)	528	513	489	466	464	459	536	517	517
T5(220mm)	553	538	510	476	476	476	557	538	544

4. Radiative transfer model and spectrum reconstruction

For a more sophisticated analysis, the line of sight observed by the spectrometer and the camera was considered and a tentative reconstruction of the spectrum was done. The corresponding radiative transfer model has to take into account radiation from the flame and from the wall. This latter contribution involves emission, but also reflection of the radiation emitted by the flame which is reflected and then partly attenuated during its second path through the flame. A multi-layer model was built following the flame shape observed with tomography and involving: 1) an atmospheric layer between the measurement setup and the flame, 2) a first cold layer of an air-methane mixture injected from the blower, 3) the high temperature region of the combustion products, 4) a second cold layer of air-methane, and the wall itself. Considering the reflection by the wall, each layer emits radiation which is partly redirected through all the successive layers. Hence eight layers are finally considered in a model involving the four layers (1 to 4) cited above and four virtual layers (5 to 8) featuring the emission toward the plate, reflected, and thus weighted by the wall reflectivity. Following a formulation similar to [6], this yields:

$$I = \sum_{j=5}^{8}[\tau(s_j,s) - \tau(s_{j-1},s)]I_0(s_{j-1}) + \rho\left\{\sum_{j=1}^{4}[\tau(s_j,s) - \tau(s_{j-1},s)]I_0(s_{j-1})\right\} + \varepsilon I_0(T_{wall})\tau(s_{wall},s) \quad (2)$$

where $\tau(s_j,s)$ stands for the transmission function of the gas layer between position s_j and the position of the spectrometer s, ρ and ε are the wall reflectivity and emissivity. The subscript ν is omitted but the intensity is of course computed for each spectral band.

Gas radiative properties are estimated using a statistical narrow band (25 cm^{-1}) model based on Malkmus' distribution of linestrengths. Model parameters, developed at CETHIL, were adjusted on LBL curves-of-growth (transmissivities as a function of the path in the gas) following the method described in Ref. [9]. Reference LBL data were generated with the CDSD-4000 spectroscopic database [10] for CO_2 and HITEMP2010 [11] for H_2O and CO. More details about those calculations can be found in Refs. [12-13]. The treatment of non-uniformities involves the Curtis-Godson approximation [14]. This approach is known to raise possible problems of over-correlation in non-isothermal conditions, but it is easy to handle and chosen in the present so as to allow a first evaluation of the method.

Spectra are built using given inlet properties (gas concentrations, flame temperature, wall temperature) and an optimized set of parameters is sought, such as to provide the best agreement between experimental data and numerical spectrum. A PSO algorithm is used for this purpose (details about the identification method may be found in [15]).

4.1 Validation of the model identification.

A first test was conducted on a synthetic spectrum built with our radiative transfer model used in a direct way. The spectrum was computed based on the following temperatures: 400 K for the wall and 1400 K for the flame, the following concentrations: 20% for H_2O, 10% for CO_2 and 5% for CO, and the respective layer thicknesses: 88.1 cm for the atmospheric layer, 5.9 cm for the first methane plus air layer, 4.4 cm for the combustion product layer and 1.7 cm for the methane plus air layer along the wall (values taken from the experiment when the line of sight is in front of the T2 position, 100 mm above the leading edge of the plate). The wall emissivity was set at the constant 0.95 value. The spectrum was then used as a mock experimental dataset and the identification algorithm was run several times with the following parameters: number of particles 20, number of loops 200.

Figure 5a shows the obtained spectra (original data and five identified spectra were all found perfectly super-imposed, showing the perfect behavior of the algorithm in this case of synthetic input data. Figure 5b shows a typical convergence criterion evolution based on the quadratic difference between both datasets as a function of the loop number.

It is known that such a test is not fully convincing because the same model is used in direct and inverse ways. However it shows that the numerical parameters and the identification algorithm are at least performing well. The wall temperature was found with an accuracy better than 1°C for all runs and all the flame properties (temperature and gas concentrations) were found with a very satisfying accuracy (for example for the first run: 1399 K for the flame with the following concentrations: 10.2% for H_2O, 20.1% for CO_2). Of course the estimations could be even further improved increasing the identification parameters (numbers of particles and loops).

Journal of Physics: Conference Series **676** (2016) 012018 doi:10.1088/1742-6596/676/1/012018

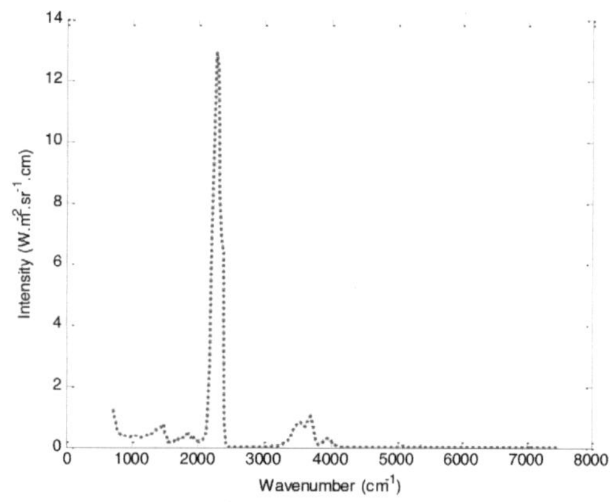

Figure 5a: synthetic spectrum (dashed line) and identified spectrum (continuous line)

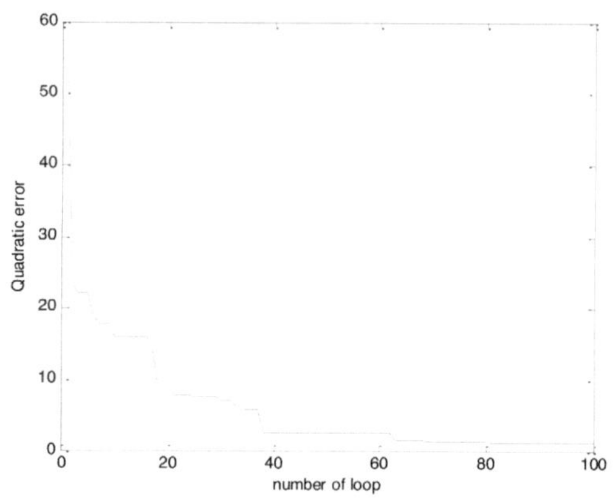

Figure 5b: convergence in terms of quadratic error between original and reconstructed data sets

4.2 Sensitivity study.

A sensitivity study was carried out still with the same method (generation of synthetic data and blind search for optimal conditions yielding the best fit in terms of spectra). Eight different cases were tested which were defined as to bracket the reference case seen above, always keeping the same geometry. The flame properties (Figure 6a), then the wall properties (Figure 6b for the wall emissivity and Figure 6c for the wall temperature), were varied in order to investigate if some conditions could penalize the ability to identify the wall temperature and the flame characteristics. The emissivity influence is plotted with a particular scale on the Y-axis (Figure 6b) since the influence of this parameter is very weak (as it even hardly appears despite the chosen scale). This also shows that this parameter has no major effect on the spectra. In all subfigures synthetic data are plotted with thin continuous lines and identified spectra are plotted with thick dotted lines, such that it appears that the spectra are apparently perfectly reconstructed since no discrepancy appear between two curves (synthetic and identified) corresponding to the same case study.

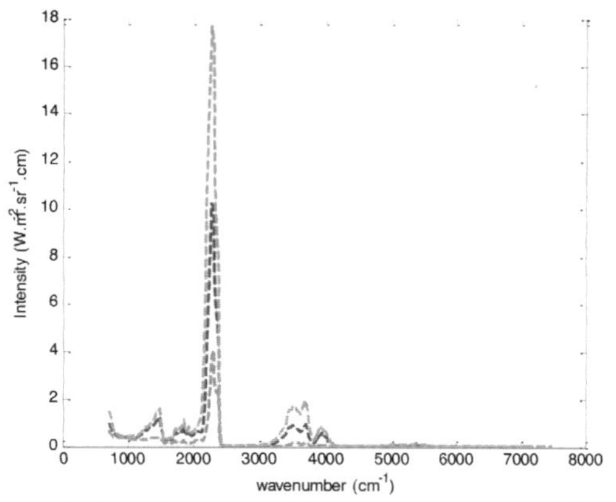

Figure 6a: Effect of the flame properties (temperature, H_2O, CO_2 and CO concentrations) set to (1000 K, 10%, 10%, 5%), (1400 K, 20%, 10%, 5%) and (1600 K, 20%, 20%, 5%) from the lowest to the highest spectrum. Wall temperature set to 400 K, emissivity 0.95.

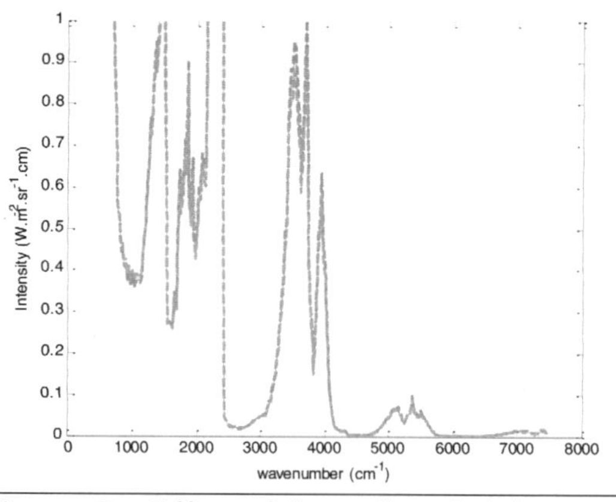

Figure 6b: Effect of the wall emissivity: set to 0.90, 0.95 and 1 from the lowest to the highest spectrum. Wall temperature set to 400 K, flame at 1400 K, with concentrations 20%, 10% and 5% for H_2O, CO_2 and CO (warning: change in the ordinate scale, because of the weakness of the effect).

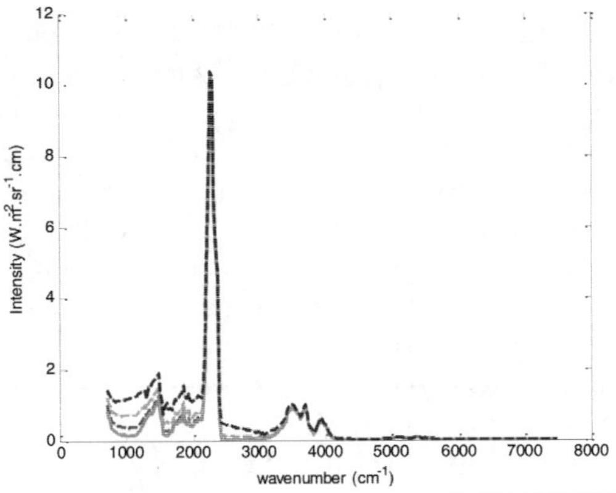

Figure 6c: Effect of the wall temperature: 300 K, 400 K, 500 K, 600 K from the lowest to the highest spectrum. Emissivity 0.95, flame at 1400 K, with concentrations 20%, 10% and 5% for H_2O, CO_2 and CO.

Figure 6: sensitivity study on the identification algorithm, comparison between synthetic data (thin continuous lines) and identified spectra (thick dotted lines, in all cases superimposed on the corresponding continuous lines)

What was especially satisfactory in this numerical exercise is that the maximum discrepancy on the temperature identification never exceeded 1 K as compared to the original value, whatever the case, either on the wall or the flame temperature. Similarly, synthetic and identified concentrations were found in a very good agreement (always below 1% of absolute discrepancy).

4.3 Results for experimental data treatment and discussion.

The method based on spectrum reconstruction and property identification was then applied on the measured spectra. The experimental data were first averaged on 25 cm^{-1} bandwidths, in accordance with the model discretization. The case of the turbulent mixing at 4% was used, the same conditions being kept as in the validation case in terms of geometry and of numerical parameters used in the PSO algorithm. The comparison between experimental data and a typical identified spectrum is presented in Figure 7a.

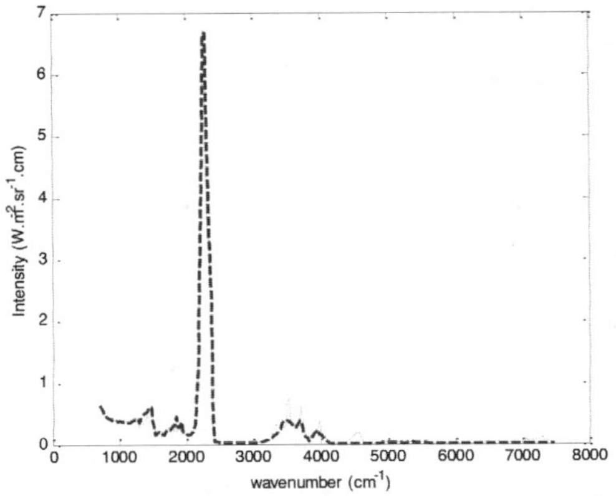

Figure 7a: Experimental dataset and reconstructed spectrum based on the search for optimal flame properties.

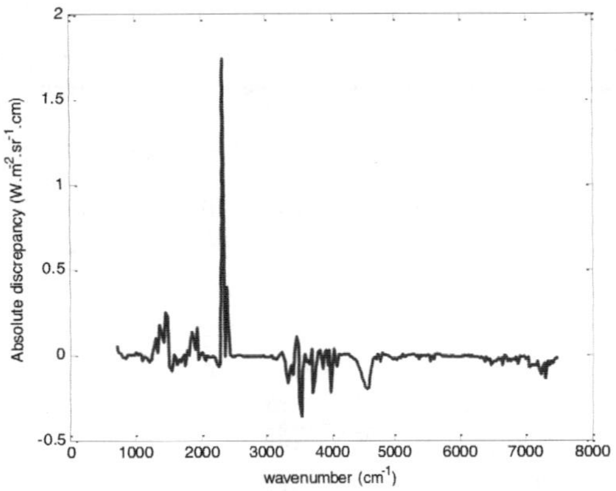

Figure 7b: Absolute discrepancy between experimental and reconstructed spectrum

It seems that the spectrum is qualitatively reconstructed, but with some discrepancies in the peaks attributed to the gases. The discrepancies are plotted in Figure 7b, revealing quite large errors which were not so obvious on the spectrum itself. Such discrepancy does not allow an identification of the flame in confidence. The identification exercise was repeated several times, also varying the number of loops in the PSO algorithm (up to 400 loops, with apparent convergence reached considering the quadratic error between experimental and numerical spectra). It was found that the identified temperature and gas concentrations were slightly varying, but an average prediction could be done on the flame properties. Spectra are not plotted because they provide exactly the same figure as Figure 7a. The average discrepancy between two identified spectra was typically below 0.01 $W.m^{-2}.sr^{-1}.cm$ (when averaged over the whole spectral range), with a maximum local value between the two extreme identifications around 0.2 $W.m^{-2}.sr^{-1}.cm$ (found in the CO_2 peak).
Regarding the identified properties, average data based on five runs were found as follows (with maximum discrepancy as compared to the average value indicated between parenthesis): temperature 1617 K (+/- 41 K), H_2O concentration 5.3% (+/-0.4%), CO_2 concentration 3.3% (+/-0.2%). CO concentration was apparently very low (around 0.2%). The wall temperature was found with less uncertainty: average value 409 K (+/- 5K). This is in a correct agreement with the other estimations, as presented above. For the flame temperature however, 1617 K is probably an overestimated value, owing to the experience of present authors involved in such flame study. The problem is that with the present model a compensating effect may affect the results since a too high temperature may be compensated by too low concentrations. At this stage we consider that the method could work well but improvements are required, in particular in terms of gas radiation modeling. A specific model, combining results from the Generalized *k*-moment method [16] together with the Multi-Spectral Framework [12] is presently developed at CETHIL for this task. Those methods were studied separately and found to provide very high accuracies when compared to LBL reference calculations both in uniform and non-uniform media. Implementing this more accurate model is likely to improve the accuracy of the outputs (in terms of temperature and species concentrations) of the inversion scheme.

As the wall temperature was the main parameter sought here and since it was obtained with less uncertainty, a systematic identification was performed on all cases, yielding the values reported in table 2. As can be seen, the agreement with the thermocouple measurements is not as good as the one observed with the data extracted from the experimental spectra and IR images, but results are still in the same range. Identified temperatures are lower as compared to the straightforward estimation (table 1). It seems that the discrepancies in the gas peaks due to the limitations of the model result in compensating deviations for the continuous background attributed to the wall. This is because the optimization is converging toward a compromise which minimizes the discrepancies over the whole wavenumber range, with overestimations and underestimations depending on the wavenumber.

Table 2. Wall temperature (in kelvins) at different positions based on complete spectrum reconstruction compared with data from micro-thermocouples, for 4 positions (x-position in mm) on the wall and 3 flow conditions of the premixing of air and methane.

	laminar		turbulent 4%		turbulent 8%	
	Identification	TC	Identification	TC	Identification	TC
T2(100 mm)	469	444	409	418	446	463
T3(140 mm)	492	475	435	447	478	500
T4(180 mm)	503	489	448	459	495	517
T5(220 mm)	523	510	459	476	512	544

Hence, the method seems to be a good idea but some refinements of the radiation model will be necessary. In particular, the model will have to address in a better way the heterogeneities in terms of concentrations and temperature.

5 Conclusion

An identification method of wall temperature and flame properties was tested on a V-shape flame, based on the post-processing of radiative transfer measurements. The idea was to reconstruct the best spectrum fitting the experimental data obtained on a real experiment and to identify the required experimental parameters. The method was shown to perform well on synthetic data produced with a SNB model for the gas radiation (also involved in the identification method). However, this must be considered with limited confidence, since this calibration is what is called an "inverse crime" because the same model is used for the data generation and the inverse identification. This is of course easier than working on real data with their noise, uncertainties and particular spectral resolution. When applied on real data, the reconstructed spectra showed a satisfying global shape but with significant discrepancies in the gas peak ranges. This was suspected to penalize the identification of the flame properties. The wall temperature was found with a better confidence and various evaluations were found in a satisfying agreement, owing to the measurement difficulties with the different methods: thermocouples, IR imaging or spectrometry. The goal of this collaborative work is still to go toward a complete flame characterization. This will require an improved gas radiation model in particular.

References

[1] C. Galizzi and D. Escudié. Experimental analysis of an oblique turbulent flame front propagating in a stratified flow, Combustion and Flame, 157, pp. 2277-2285, 2010.

[2] H. Guo, B. Tayebi, C. Galizzi and D. Escudié. Burning rates and surface characteristics of hydrogen-enriched turbulent lean premixed methane-air flames, International Journal of Hydrogen Energy, 35, pp. 11342-11348, 2010.

[3] Y. Billaud, P. Boulet, Y. Pizzo, G. Parent, Z. Acem, A. Kaiss, A. Collin and B. Porterie Determination of woody fuel flame properties by means of emission spectroscopy using a genetic algorithm. Combustion Science and Technology, 185, pp. 579-599, 2012.

[4] A. Soufiani, J.-P. Martin, J.-C. Rolon and L. Brenez. Sensitivity of temperature and concentration measurements in hot gases from FTIR emission spectroscopy. Journal of Quantitative Spectroscopy and Radiative Transfer, 73, pp. 317-327, 2002.

[5] F. André, R. Vaillon, I. Ayranci, D. Escudie and N. Selcuk. Radiative transfer diagnostic technique of sooting flames frome mission spectroscopy. In: Proceedings of the Fourth International Symposium on Radiative Transfer, Istambul, Turkey, June 20-25, 2004.

[6] P. Vena, C. Galizzi, F. André, M. Kuhni and D. Escudié. Experimental study of flame-wall interaction: focus on the propagation of a premixed front near a wall, 23rd International Symposium on Transport Phenomena, Auckland, New Zealand, 19-22 November 2012.

[7] G. Parent, Z. Acem, S. Lechêne and P. Boulet. Measurement of infrared radiation emitted by the flame of a vegetation fire. International Journal of Thermal Sciences, 49, pp. 555-562, 2010.

[8] P. Boulet, G. Parent, Z. Acem, B. Porterie, A. Kaiss, Y. Billaud, Y. Pizzo and C. Picard. Experimental investigation of radiation emitted by optically-thin to optically-thick wildland flames Journal of Combustion, vol. 2011, Article ID 137437.

[9] J. Taine and A. Soufiani. Gas IR radiative properties: from spectroscopic data to approximate model. Advances in Heat Transfer, 33, pp. 295-414, 1999.

[10] S.A. Tashkun and V.I. Perevalov. CDSD-4000: High-resolution, high-temperature carbon dioxide spectroscopic databank. Journal of Quantitative Spectroscopy and Radiative Transfer, 112, pp. 1403-1410, 2011.

[11] L.S. Rothman et al. HITEMP, the high-temperature molecular spectroscopic database. Journal of Quantitative Spectroscopy and Radiative Transfer, 111, pp. 2139-2150, 2010.

[12] F. André, L. Hou, M. Roger and R. Vaillon. The multispectral gas radiation modeling: a new theoretical framework based on a multidimensional approach to *k*-distribution methods, Journal of Quantitative Spectroscopy and Radiative Transfer, 147, pp. 178-195, 2014.

[13] F. André, L. Hou and V.P. Solovjov. An exact formulation of k-distribution methods in non-uniform gaseous media and its approximate treatment within the Multi-Spectral framework, accepted in the proceedings of the present Eurotherm seminar, 2015.

[14] S.J. Young. Nonisothermal band model theory. JQSRT Vol. 18, 1977, pp. 1-28.

[15] M. Clerc. Particle Swarm Optimization. ISTE Publishing Company, London, UK, 2006.

[16] F. André, V.P. Solovjov, L. Hou, R. Vaillon and D. Lemonnier. The generalized k-moment method for the modeling of cumulative k-distributions of H2O at high temperature. Journal of Quantitative Spectroscopy and Radiative Transfer, 143, pp. 92-99, 2014.

Journal of Physics: Conference Series **676** (2016) 012020 doi:10.1088/1742-6596/676/1/012020

Infrared signature modelling of a rocket jet plume – comparison with flight measurements

V Rialland[1], A Guy[2], D Gueyffier[2], P Perez[1], A Roblin[1], T Smithson[3]

[1] ONERA DOTA, BP 80100, 91123 Palaiseau CEDEX, France
[2] ONERA DEFA, BP 80100, 91123 Palaiseau CEDEX, France
[3] DRDC Valcartier, 2459 rue de la Bravoure, Québec, Québec, G3J 1X5, Canada

Corresponding author: Valerie.Rialland@onera.fr

Abstract. The infrared signature modelling of rocket plumes is a challenging problem involving rocket geometry, propellant composition, combustion modelling, trajectory calculations, fluid mechanics, atmosphere modelling, calculation of gas and particles radiative properties and of radiative transfer through the atmosphere. This paper presents ONERA simulation tools chained together to achieve infrared signature prediction, and the comparison of the estimated and measured signatures of an in-flight rocket plume. We consider the case of a solid rocket motor with aluminized propellant, the Black Brant sounding rocket. The calculation case reproduces the conditions of an experimental rocket launch, performed at White Sands in 1997, for which we obtained high quality infrared signature data sets from DRDC Valcartier. The jet plume is calculated using an in-house CFD software called CEDRE. The plume infrared signature is then computed on the spectral interval 1900-5000 cm^{-1} with a step of 5 cm^{-1}. The models and their hypotheses are presented and discussed. Then the resulting plume properties, radiance and spectra are detailed. Finally, the estimated infrared signature is compared with the spectral imaging measurements. The discrepancies are analyzed and discussed.

1. Introduction

Infrared space or airborne surveillance sensors are the most effective systems to detect and track proliferating ballistic missiles [1]. These sensors have to be designed and optimized upon realistic infrared signatures of plume rockets. The lack of observations in operational conditions leads to the use of simulations to complete ground measurements. The infrared signature modeling of rocket plumes is a challenging problem involving rocket geometry, propellant composition, combustion modeling, trajectory calculations, fluid mechanics, atmosphere modeling, gas and particles radiative properties, and radiative transfer through the atmosphere. ONERA decided to tackle this complex problem through a common project involving several research teams. This paper presents the simulation tools chained together to achieve the infrared signature prediction and the comparison of the estimated and measured signatures of an in-flight rocket plume.

We considered the case of a solid rocket motor with aluminized propellant, the Black Brant sounding rocket (maximum thrust 77 kN). The calculation case reproduces the conditions of an experimental rocket launch, performed at White Sands in 1997, during which DRDC Valcartier

Journal of Physics: Conference Series **676** (2016) 012020 doi:10.1088/1742-6596/676/1/012020

recorded high quality infrared signature data. Records were collected by a spectral imager of 8 by 8 elements, associated with a tracking system, with a spectral resolution of 4 cm^{-1} on the spectral interval 1900-5000 cm^{-1} and a time interval of about 420 ms. The Black Brant was the second stage of the sounding rocket. The measurements encompass 51 frames acquired over 20 seconds of the Black Brant propulsion phase, from altitude 5 km to 23 km. The distance between the observer and the missile launch site is about 13.5 km. A test point is chosen at altitude 8 km. This point was selected as our first calculation point in order to limit the effects of high altitude on the plume physics and because it corresponds to an acquisition where the whole plume is visible in the frame. At this altitude, the rocket is in a stabilized phase of thrust.

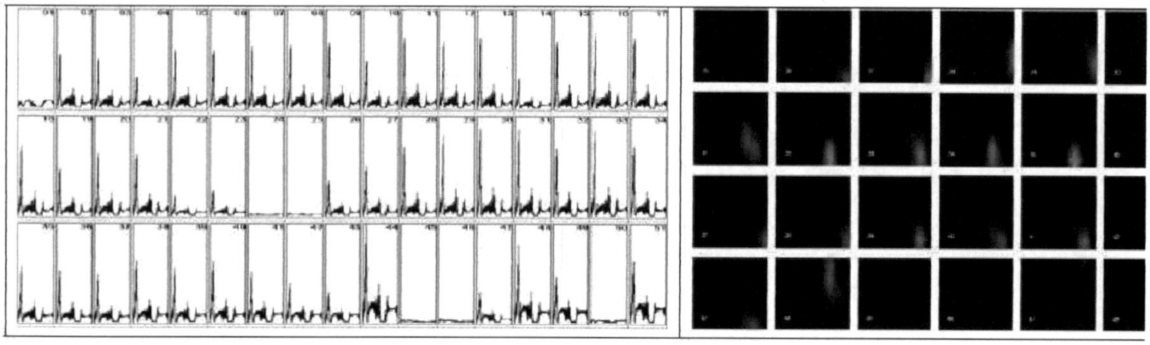

Figure 1: Examples of DRDC Valcartier spectral imager measurements for Black Brant launch in White Sands (1997)

2. Models and hypotheses

2.1. Restitution

A preliminary work (not detailed here) was conducted in order to retrieve the propulsion characteristics of the sounding rocket Black Brant launched on November 18th 1997 at White Sands Missile Range. The first stage (not studied) was a Terrier MK 70, while the second stage was a Black Brant VC MOD2 [2][3]. Insights about propellant composition and shape, nozzle geometry and propulsion elements were inferred from references [4][5][6].

2.2. Combustion and CFD models

The modeling of the infrared signature (IRS) of the Black Brant plume is obtained using a chain of multi-physics codes developed at ONERA. The computational flow starts from the combustion chamber and nozzle properties, whose results are used both for the trajectory calculation and for the plume flowfield computation, and ends with the radiative transfer computation to retrieve the infrared signature as seen by the spectro-imaging sensor.

COPPELIA is a thermodynamic equilibrium code that computes the motor specific impulse, the temperature and the gaseous species composition in the combustion chamber as a function of the propellant composition and the nozzle expansion ratio, using either a frozen or equilibrium expansion model. The propellant composition retained for the Black Brant calculation is: 19% Al, 61% of ammonium perchlorate, 20% polyurethane. COPPELIA outputs are used to compute the Black Brant trajectory in order to retrieve the rocket speed, the distance from the sensor to the target and the aspect angle of the plume with respect to the sensor line of sight. The chamber temperature, pressure and gaseous composition, the rocket speed, altitude and geometry are used to initialize ONERA computational fluid dynamics code CEDRE for the plume flow field computation.

Table 1: Input characteristics of the Black Brant 8km test case

Altitude (m)					7900					
Distance from sensor (m)					14221					
Aspect angle (deg)					112.4					
Speed (m/s)					646.8					
Flight Mach Number					2.11					
Atmospheric pressure (Pa)					36120					
Atmospheric temperature (K)					236.8					
Chamber pressure (bar)					45.2					
Chamber temperature (K)					3290.8					
Nozzle throat area (m²)					0.0111					
Nozzle exit area (m²)					0.0871					
Mass fraction at the nozzle entrance (%)										
H	O	OH	H_2	H_2O	CO	CO_2	HCl	CL	N_2	Al_2O_3
0,12%	0,01%	0,22%	2,88%	6,59%	28,95%	1,52%	16,45%	0,93%	7,98%	34,34%

CEDRE [7] is a multiphysics computational tool for numerical simulation in the field of energetics, with particular emphasis on propulsion applications. The code can handle several coupled physical subsystems, each of them being taken into account by a specialized time-dependent solver. The compressible flow module solves the Navier-Stokes equations with any number of species and various possible models for chemical reactions. The simulation space can be 3D or 2D (plane or axisymmetric). For turbulence, RANS 2- or 4-equations models (k-ε or k-ω...) are available, as well as Large-Eddy Simulation (in the MILES approach or with Smagorinsky subgrid models) [8]. Liquid and solid particles can be modelled using a Lagrangian [9] or Eulerian [10] approach depending on applications. With both particle solvers, interactions with the gas include drag forces, heat exchanges and vaporization-condensation.

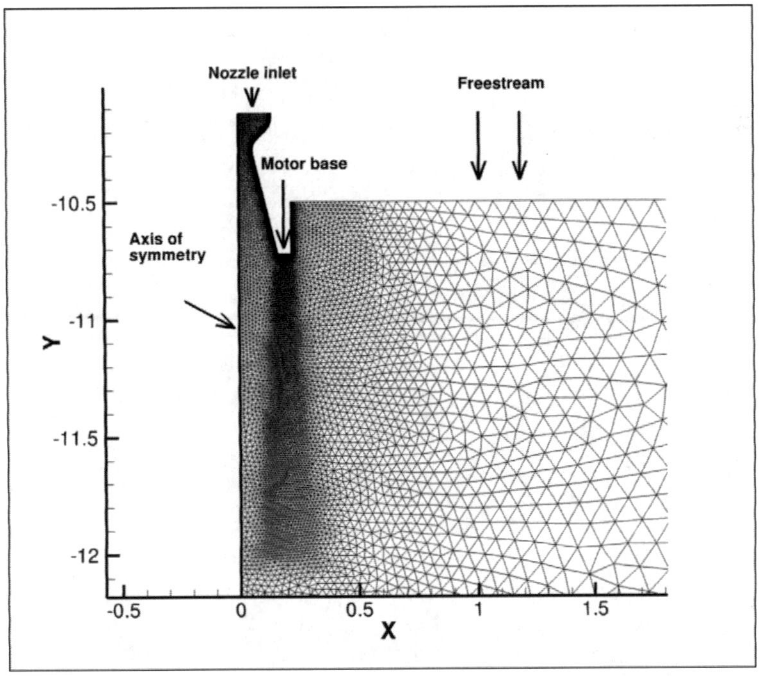

Figure 2: Axisymmetric non structured mesh used for the CFD computation (close-up in nozzle region)

For the Black Brant test case we have conducted axisymmetric calculations of the flow on an unstructured mesh (500m x 10m domain, 191K cells) refined on boundaries and in the mixing layer (figure 2). The HLLC scheme with 2nd order reconstruction along with the VanLeer slope limiter is used for space discretization of the convective terms. Time integration is performed using a one-step Euler implicit scheme until a steady-state is obtained. A RANS k-ε turbulence approach is used. However, these Reynolds-averaged Navier-Stokes models are not fully adapted to axisymmetric flows. Therefore, some constants of the original model by Menter [11] were modified following a strategy already used in [12][13]. The chemistry is modeled using a semi-detailed kinetic scheme shown in Table 2. The chemical scheme consists of three main sets of reactions: H_2 and O_2 reactions 1 to 8, CO and CO_2 reactions 9 to 11 and chlorine species reactions 12 to 17. The alumina particles are tracked by the Eulerian dispersed phase solver coupled with the Navier-Stokes solver, and are assumed to remain liquid. No coupling between chemistry and turbulence is considered and radiation is not taken into account in the enthalpy balance.

Having no information about particle size distribution, we resorted to Hermsen's correlation [14] which relates the particle mass weighted average diameter D43 to the rocket nozzle diameter. We assumed that there was only one class of particles with diameter 5.71 μm (every particle in the plume has the same diameter). The particle dispersion is modeled with the Eulerian approach. The complex refractive index of alumina is computed according to Anfimov empirical correlation [15].

Table 2: Chemical reactions taken into account in the CFD simulation

Number	Reaction
1	$O_2 + H \leftrightarrow O + OH$
2	$H_2 + O \leftrightarrow H + OH$
3	$H_2 + OH \leftrightarrow H_2O + H$
4	$2\,OH \leftrightarrow H_2O + O$
5	$2\,H + M \leftrightarrow H_2 + M$
6	$H + OH + M \leftrightarrow H_2O + M$
7	$H + O + M \leftrightarrow OH + M$
8	$2\,O + M \leftrightarrow O_2 + M$
9	$OH + CO \leftrightarrow H + CO_2$
10	$O_2 + CO \leftrightarrow O + CO_2$
11	$O + CO + M \leftrightarrow CO_2 + M$
12	$H + HCl \leftrightarrow H_2 + Cl$
13	$H + Cl_2 \leftrightarrow HCl + Cl$
14	$OH + HCl \leftrightarrow H_2O + Cl$
15	$O + HCl \leftrightarrow OH + Cl$
16	$2\,Cl + M \leftrightarrow Cl_2 + M$
17	$H + Cl + M \leftrightarrow HCl + M$

2.3. Radiative transfer models

The aerothermochemical cartography of the plume computed by CEDRE, which is detailed in the next paragraph, is transposed on a structured mesh for the last step of the modeling chain: the radiative transfer computation. Our infrared signature computation tool, named SIR, takes as input the pressure, temperature, mass fraction of H_2O, CO_2, CO, HCl, alumina particle density and temperature. It can perform radiative transfer computation in a non-scattering medium with either a line-by-line model or a statistical narrow band model called RGM3000. The scattering media are dealt with using the SHDOM solver described below.

RGM3000 is based on Curtis-Godson approximation [16], which takes into account the spectral correlations in case of multiple layers of gas with different physical conditions. Voigt line profiles inside a narrow band are also considered using the empirical formulation of Crisp et al [17], where a Doppler mean broadening [18] in a narrow band is combined to the Lorentz mean broadening. In rocket plume detection situations, where radiation emitted by hot gases is seen through kilometers of cold atmosphere, the Curtis-Godson approximation is not sufficient to treat the temperature and concentration gradients, and it is known to lead to large errors. For such cases, Ludwig et al [19] suggested to separate spectral lines inside a narrow band into different groups. The assumption is made that lines belonging to different groups are uncorrelated and the Curtis-Godson approximation can be applied to each line group. All the lines of a group are considered to belong to a fictitious gas and the original gas is then a mixture of these independent fictitious gases. The concept of fictitious gases has been discussed and applied to cumulative-k models [20][21] and to statistical models [22]. It has been implemented in our infrared signature code SIR with 5 fictitious gases for H_2O and CO_2. Particles are not taken into account yet in this model, neither in absorption nor in emission.

The Black Brant plume also contains liquid or solid alumina (Al_2O_3) particles. These particles strongly emit in the infrared region of the spectrum. It forces us to consider the radiation scattering phenomenon as an important part of the model. Computing the RTE solution with scattering along a given line of sight (LOS) involves the computation of multiple other LOS whose contribution is scattered into the first LOS direction. As a consequence, there is no simple analytical solution to the RTE that can be spectrally averaged. The inadequacy between scattering and spectrally-averaged gas absorption description has been underlined by several authors since decades [23][24][25].

The radiative transfer equation solver named SHDOM (Spherical Harmonics Discrete Ordinates Method) has been developed by Evans [26] in order to study the influence of the heterogeneities of the clouds in the transmission and scattering of radiation. Since geometrical problems, such as boundary conditions, are similar between atmospheric clouds and exhaust plumes, it was decided to use this solver in our application, with modifications consisting in taking into account high temperature combustion gas spectral data (and alumina particles optical properties for further calculations). It has to be noted that SHDOM has been already used to compute ultraviolet emission from alumina loaded rocket exhaust plume [27]. The SHDOM solver discretizes the radiance field by representing the source function at grid points with an adaptive spherical harmonics expansion in the angular variables. The solution method is iterative and consists of four steps. First, the spherical harmonics source function is transformed into discrete ordinates. Then, the source function is integrated to obtain the radiance field. The integral form of the radiative transfer equation (RTE) grid (1) is solved at each point of the field:

$$I^n_{uv}(s) = \exp\left[-\int_0^s K_{ext}(s')ds'\right] I^n_{uv}(0) + \int_0^s \exp\left[-\int_{s'}^s K_{ext}(t)dt\right] J^{n-1}_{uv}(s') \cdot K_{ext}(s')ds' , \quad (1)$$

where I is the radiance, J the source function and K_{ext}, is the extinction coefficient. The subscripts u and v stand respectively for the zenithal angle and the azimuth angle, and the superscript n indicates the iteration step.

In the third step, the radiance field is transformed back to spherical harmonics. The source function is then computed from the radiance field in spherical harmonics. The radiance and source function are initialized with an Eddington radiative transfer solution on independent columns of the base grid.

The value for K_{ext} is determined for each gas cell by calculating the mean transmissivity of the cell in one direction by the Malkmus function for each narrow band:

$$\bar{\tau}(l) = \exp\left[\phi\left(1-\left(1+\frac{2\bar{\kappa}l}{\phi}\right)^{1/2}\right)\right] \qquad \phi = 2\frac{\bar{\gamma}}{\bar{\delta}} \quad , (2)$$

where $\overline{\tau}(l)$ is the mean transmissivity on a narrow band, for gas layer of thickness l, κ, $\overline{\gamma}$, $\overline{\delta}$ are respectively the mean absorption coefficient, the mean half-width, and the mean line spacing in the narrow band. K_{ext} is then obtained simply by:

$$K_{ext} = -\frac{\log(\overline{\tau}(l))}{l} \quad . (3)$$

The choice for the length l depends on each cell and on the configuration of the medium. For a plane parallel medium, l is simply the height of each layer. This approach is retained for further complete calculations on a jet plume containing gas and particles because of his ability to easily take scattering into account [28]. But it is not able to take spectral correlations into account.

For scattering media, SIR computes the medium optical properties: gas and particle temperatures, phase function, extinction coefficient (with formula (2)) and albedo for every cell. The particle temperature is different from the gas temperature because particles have higher thermal inertia. Their temperature also depends on their diameter. The extinction coefficient for particles and the phase function are computed according to Mie theory. The phase function is given with Henyey Greenstein formula. These data are sent to SHDOM. SHDOM outputs the spectral radiance over a 2D grid that samples the axisymmetric plume for the user-specified aspect angles. SIR performs the spectral and spatial integration to obtain the source intensity in the required spectral band. The infrared signature is computed with no transmission on the spectral interval 1900-5000 cm^{-1} using a step of 5 cm^{-1}, on 2D grid with 0.2x0.2 m² pixels, with an aspect angle of 112° corresponding to the measurements. The atmospheric transmission from the target to the sensor is computed using the US Standard model and multiplied with the plume spectrum.

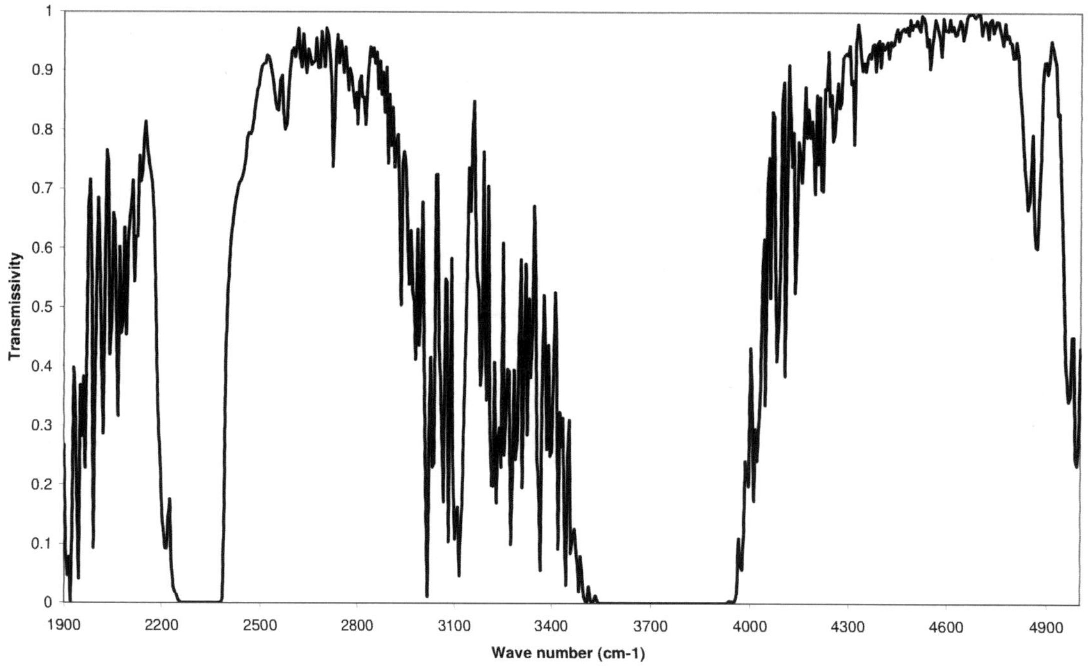

Figure 3: Atmospheric transmission from the plume to the sensor

Journal of Physics: Conference Series **676** (2016) 012020 doi:10.1088/1742-6596/676/1/012020

3. Simulation results

Some of CEDRE outputs are displayed on figure 4. The Mach number plot shows the potential core of the round jet, which corresponds to the red area on the top left side. The plume length is about 300m long whereas the flame, which can be seen on the OH mass fraction plot, is only 40m long. The afterburning maximum is located roughly at 18m from the nozzle exit plane on the temperature plot. The particle number plot shows that particles are mostly gathered near the nozzle exit.

Figure 5 shows the evolution of temperature and radiative species mass fractions along the plume axis. A single shock cell is present in this flow, located 70 cm downstream the nozzle. Carbon monoxide is the most abundant radiative species in the nozzle exit plane. CO mass fraction then decreases as it oxides into CO_2, which is maximum near the afterburning maximum. Water vapor is the product of OH hydrogenation, and HCl turns into H_2O and Cl_2.

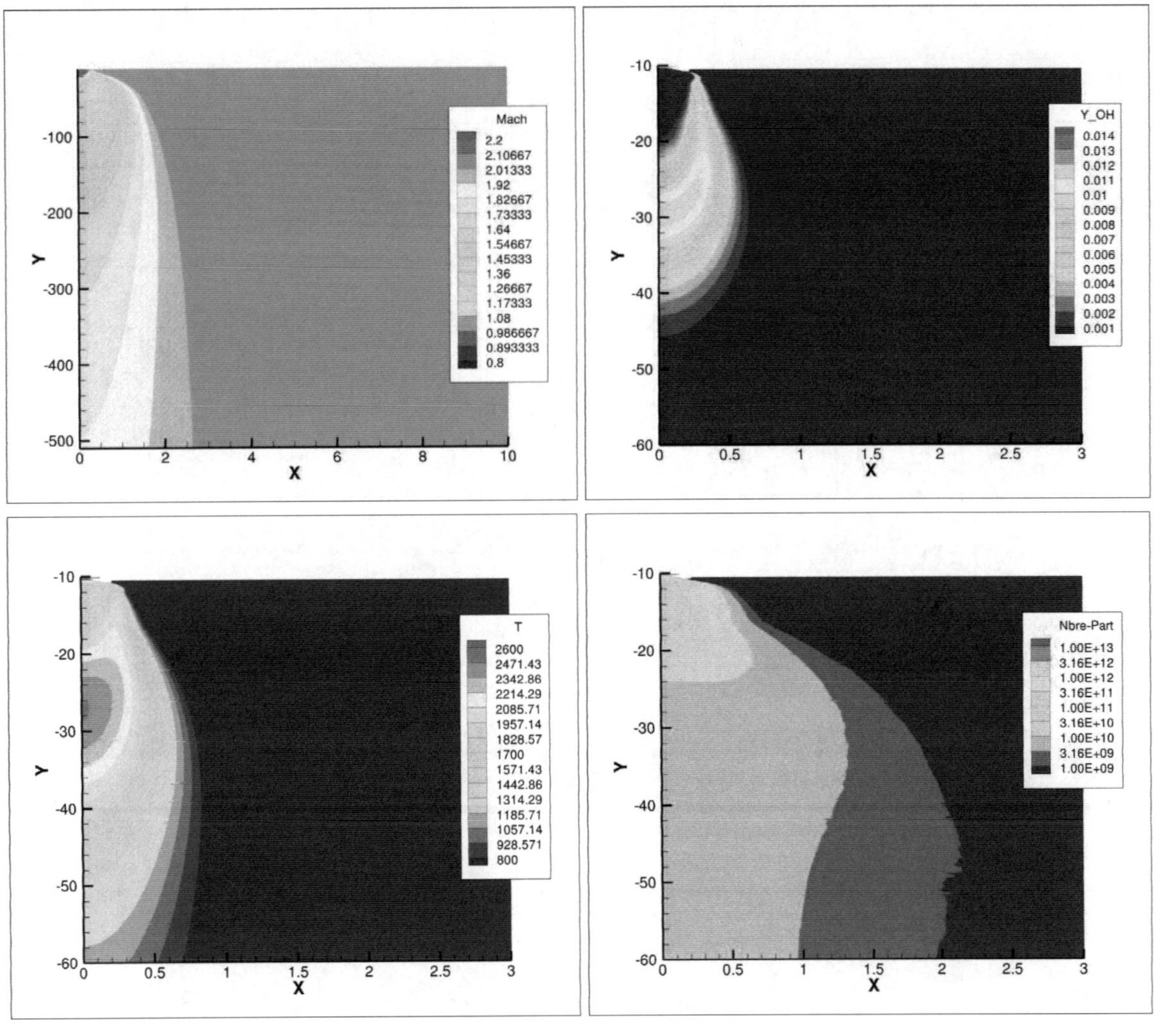

Figure 4: CEDRE outputs: Mach number, OH mass fraction, gas temperature (K), alumina particles number (m^{-3})

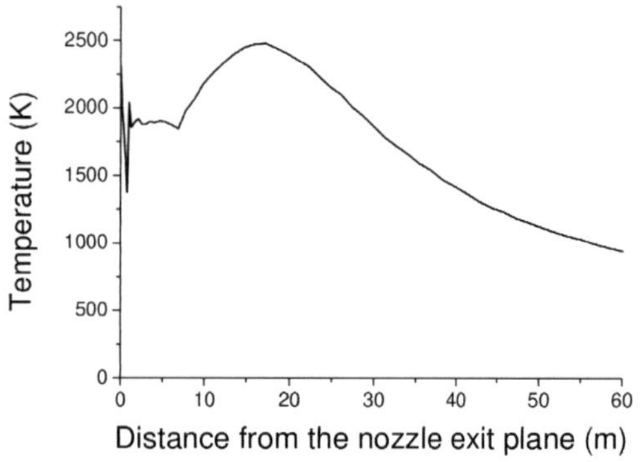

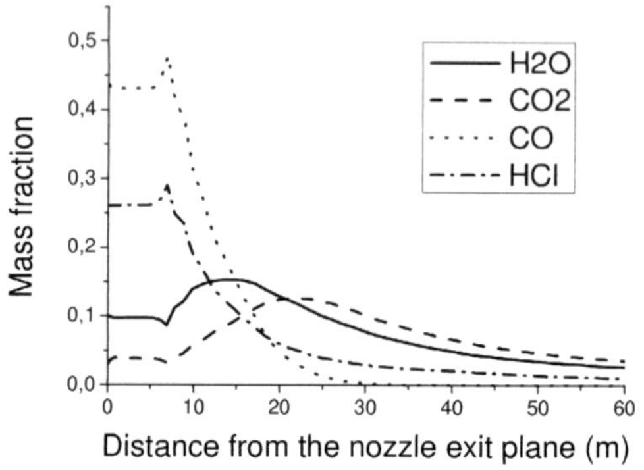

Figure 5: CEDRE outputs: gas temperature and radiative species mass fraction along the plume axis

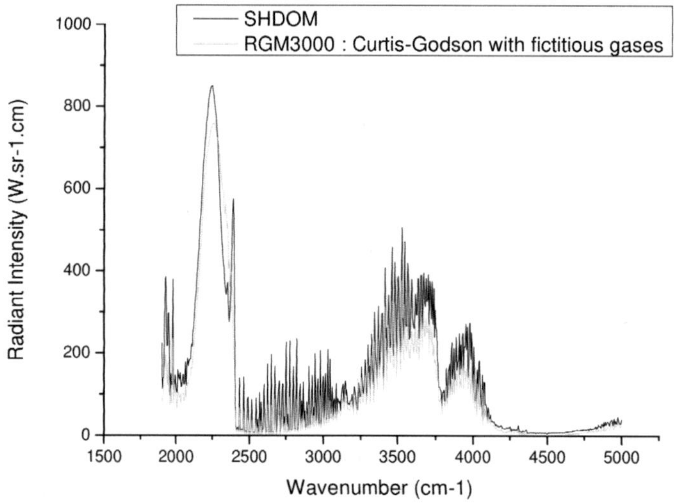

Figure 6: Black Brant spectral intensity at altitude 7.9 km, spectral resolution 5 cm^{-1}

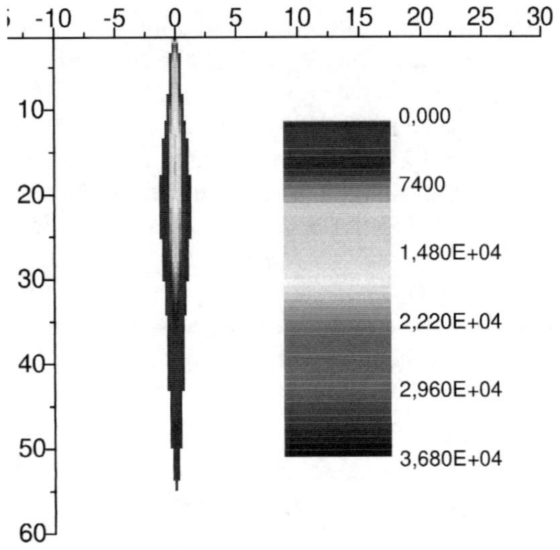

Figure 7: Black Brant 1900-5000 cm^{-1} spectrally integrated radiance cartography (pixels 0.2 m)
Units : $W.sr^{-1}.m^{-2}$

Figure 6 shows the Black Brant plume spectral intensity computed by SHDOM and RGM3000 at the source. CO emission lines can be seen on 1900-2000 cm^{-1} spectral interval. HCl emission lines on 2500-3000 cm^{-1} and water emission lines on 3250-4000 cm^{-1} are overestimated by SHDOM. This results from the uncorrelated calculation of the transmittivity as an input of SHDOM. The particle emission continuum computed by SHDOM is only slightly above the intensity computed by RGM3000 on an unscattering medium around 4500 cm^{-1}.

Figure 7 represents the radiance cartography of the Black Brant plume computed by SHDOM with an aspect angle of 112°: 0° is the front and 180° is the back of the rocket. The maximum intensity is located between 18 an 20 m from the nozzle exit plane. It coincides with the temperature maximum, and is close to H_2O and CO_2 maxima, the main radiative contributors in this large spectral band (1900-5000 cm^{-1}).

4. Comparison with spectro-imaging measurements

To perform the comparison between the spectro-imaging measurements and the simulation results, we applied to the SHDOM results for the source the atmospheric transmission from the plume to the sensor presented on Figure 3. We then computed the plume apparent intensity with RGM3000 model taking into account the atmospheric path from the sensor to the plume. Figure 8 compares the spatially integrated spectral measurements to the apparent spectral intensity computed by SHDOM and RGM3000. There is a good agreement between measurements and calculations, except on the emission line at 2200 cm^{-1}, which is slightly overestimated by RGM3000. The overestimation of HCl spectral lines by SHDOM is clearly visible on 2500-3000 cm^{-1}, whereas RGM3000 results are closer to the measurements on that spectral band. On the contrary, the particle emission continuum computed by SHDOM around 4500 cm^{-1} is closer to the measurements even if lower.

In order to take advantage of the spectro-imagery, we integrated the apparent radiance over the pixel surface, which gives an apparent intensity per pixel in $W.sr^{-1}$, then summed the pixels along the direction orthogonal to the plume axis. This spatial information gives us insight on the spectral behavior on different stations along the plume axis. Figure 9 compares the measured station radiation integrated over the whole instrument spectral band 1900-5000 cm^{-1} with SHDOM results (computed on a larger spatial domain). A very positive point is that the computed and the measured plumes have same length, which validates our choice for the RANS turbulence model used in CEDRE. One can see that the computed station radiance is close to the measured station radiance for stations farther than 20 m from the nozzle exit plane. The first two closest stations are overestimated by our simulation. Overall, we obtain a good agreement between calculations and measurements.

We can also analyze whether the discrepancies between our computations and the measurements originate either from the gas or from the particle radiation computations. On figure 10, we integrated the station radiation on spectral bands 1900-2100 cm^{-1}, which encompasses the CO_2 emitting line, and on 4400-4600 cm^{-1}, which is located in the particle emission continuum. The left graph, which shows the gas spectral band, is very similar to the whole spectral band integration represented on figure 9: the CO_2 emission close to the nozzle exit plane is overestimated in our simulations. On the right graph, which shows the particle spectral band, one can see that the particle emission continuum is underestimated by CEDRE and SHDOM, except for the first station located about 3 m from the nozzle exit plane. This may indicate that the alumina particle properties and behavior in the plume are not well modeled far from the nozzle exit.

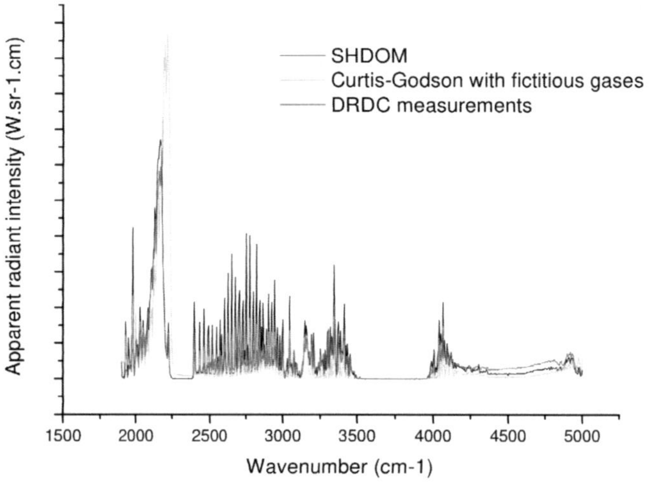

Figure 8: Black Brant apparent spectral intensity at 7.9 km as seen by the sensor
Spectral resolutions : 5 cm^{-1} for calculations, 4 cm^{-1} for measurements

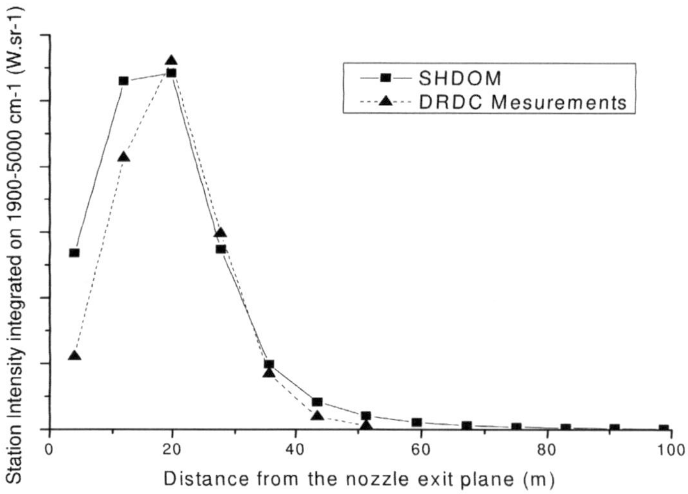

Figure 9: Plume sradiation at 7.9 km – spectral band 1900-5000 cm^{-1} (nozzle on the left side)

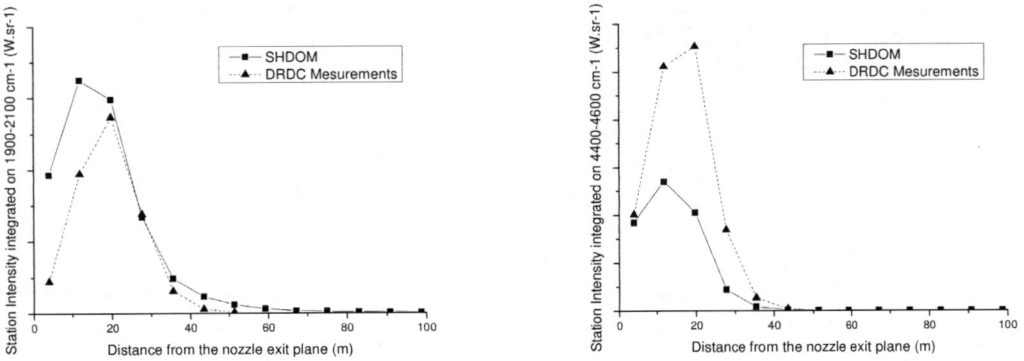

Figure 10 : Plume station radiation at 7.9 km –spectral bands 1900-2100 cm^{-1} (left), 4400-4600 cm^{-1} (right)
(Nozzle is on the left side on both graphs)

5. Conclusion

We have developed a chain of simulation tools in order to predict rocket plume infrared signatures. This chain involves multiphysics codes performing combustion computation, compressible, turbulent and multiphase flow simulation, and radiative transfer calculation. We have made use of this chain to simulate a Black Brant sounding rocket plume and our computations are compared to infrared spectral imaging measurements. The infrared signature is computed with two radiative transfer codes: the SHDOM solver which takes into account scattering on alumina particles, and RGM3000 which is a statistical narrow-band model using the Curtis-Godson approximation with fictitious gases. The simulated spectrum is close to the measured spectrum: RGM3000 is closer on gas emission lines than SHDOM, whereas SHDOM better reproduces particle emission continuum. However the particle radiation is underestimated by our simulation tools, especially in the region far from the nozzle exit. The calculated and measured plumes have the same length which validates the RANS turbulence model used in CEDRE. Given the uncertainties on the restitution of the motor parameters, on modelling, on weather and experimental conditions, we believe that we have achieved a very good agreement between simulations and measurements.

A sensitivity analysis around the 8km calculation point will be conducted in order to consolidate these first results. It will also help us finding out which modeling features we must improve, such as particle dynamics in the CEDRE solver or spectral correlations in scattering media. In particular, the Radiance Splitting Method described in [28] will be implemented in 3D scattering media. We will then address calculation points at higher altitudes and speed.

References

[1] Roblin A, Chervet P, Deschamps J, Donath T, Primot J, Herbin S and Champagnat F 2004 Engineering and Technical studies relating to an airborne infrared early warning system *2nd AAAF International Conference on Missile Defence "Challenges in Europe"* Florence (Italy)

[2] Moyers R L, et al. 1998 Ground-Based Signature Measurements of a Black Brant Sounding Rocket *Proc. NATO-IRIS* **43** No. 1

[3] Schwartz M, et al. 1999 The SERTS-97 Rocket Experiment to Study Activity on the Sun : Flight 36.167-GS on 1997 November 18 *NASA/TP-1999-208640*

[4] Smithson T, Fournier P, Turbide S 2011 Missile Exhaust Plume Phenomenology Studies Using IR Imaging Spectroscopy *Private communication*

[5] Department of the Air Force 1997 Theater Ballistic Missile Targets Programmatic

Environmental Assessment

[6] U.S. Air Force 1997 Final Environnmental Impact State for the Program Definition and Risk Reduction Phase of the Airborne Laser Program

[7] Chevalier P, Courbet B, Dutoya D, Klotz P, Ruiz E, Troyes J, and Villedieu P 2005 CEDRE : development and validation of a mulitphysic computational software *1st European conference for aerospace sciences*

[8] Bertier N, Courbet B, Dutoya D, Vuillot F and Sagaut P 2004 Large eddy simulation of a subsonic flow over a cavity on general unstructured grids *AIAA paper* 0679

[9] Villedieu P 2002 SPARTE : Modèles physiques et méthodes numériques *ONERA Technical Report RT 2/06940 DTIM*

[10] Ruiz E. 1999 Un modèle hyperbolique pour les écoulements polyphasiques dispersés en maillage non structuré généralisé *14ème Congrès Français de Mécanique* Toulouse (France)

[11] Menter F R 1994 Two-equation eddy-viscosity turbulence models for engineering applications *AIAA Journal* **32** 1598-1605

[12] Turpin G, Troyes J 2000 Validation of a twoequation turbulence model for axisymmetric reacting and non-reacting flows *36th AIAA/ASME/SAE/ASEE Joint Propulsion Conference & Exhibit* Huntsville (AL) paper 3463

[13] Gueyffier D, Fromentin-Denoziere B, Simon J, Merlen A and Giovangigli V 2014 Numerical Simulation of Ionized Rocket Plumes *Journal of Thermophysics and Heat Transfer* **28** 218-225

[14] Hermsen R W 1981 Aluminium Oxide Particle Size for Solid Rocket Motor Performance Prediction *J. of Spacecraft* **18** 483-490

[15] Anfimov N A, Karabadjak G F, Khmelinin B A, Plastinin Y A, Rodionov AV 1993 Analysis of mechanisms and nature of radiation from aluminium oxide in different phase states in solid rocket exhaust plumes *AIAA 28 th Thermophysics conference*

[16] Godson WL 1953 The evaluation of infra-red radiative fluxes due to atmospheric water vapour *Quart. J. Roy. Meteo. Soc.* **73** 367-379

[17] Crisp D, Fels SB and Schwarzkopf MD 1986 Approximate methods for finding CO2 15-μm band transmission in planetary atmopsheres *J. Geophys.. Res.* **91(D11)** 11851-11866

[18] Malkmus W 1968 Random band models with lines of pure Doppler shape *J. Opt. Soc. Am.* **58** 1214-1217

[19] Ludwig CB, Malkmus W, Reardon JE and Thompson JAL 1973 *Handbook of Infrared Radiation from Combustion Gases* (NASA)

[20] Rivière P, Soufiani A and Taine J 1995 Cumulative-k-fictitious gas model for H2O IR radiation in the Voigt regime *J. Quant. Spec. Rad. Transf.* **53** 335-346

[21] Rivière P, Soufiani A and Taine J 1995 Correlated-k and fictitious gas methods for H2O near 2.7 μm *J. Quant. Spec. Rad. Transf.* **48** 187-203

[22] Soufiani A André F and Taine J 2002 A fictitious-gas based statistical narrow-band model for IR long-range sensing of H2O at high temperature *J. Quant. Spec. Rad. Transf.* **73** 339-347

[23] Ludwig CB, Malkmus W, Freeman GN, Reed R and Slack M 1980 Infrared radiation from rocket plumes *SPIE* **253** 122-128

[24] Malkmus W X 1980 Some comments on the extension of band models to include scattering *J. Rad. Transf.* **40** 201-204

[25] Reed RA et al. 1994 Compatibility of infrared band models with scattering *J. Thermophys. Heat Transf.* **8** 208-215

[26] Evans, K. 1998 The Spherical Harmonics Discrete Ordinate Method for Three-Dimensional Atmospheric Radiative Transfer *Journal of the Atmospheric Sciences* **55** 429-446

[27] Baudoux PE 2002 Développement d'un modèle de transfert radiatif 3D adapté au calcul de la signature d'un jet de moteur-fusée dans l'ultraviolet *PhD*

[28] Pautrizel JB, Roblin A, Perez P and Rialland V 2010 Absorption-scattering coupling for the infrared signature of an aluminized solid rocket motor *ICHMT RAD-10*

Experimental validation and modelling of a photobioreactor operating with diluted and controlled light flux

V ROCHATTE[1, 3], J DAUCHET[2, 3], J-F CORNET[2, 3]

[1] Clermont Université, Université Blaise Pascal, Institut Pascal, BP 10448, F-63000 Clermont-Ferrand, France

[2] Clermont Université, ENSCCF, Institut Pascal, BP 10448, F-63000 Clermont-Ferrand, France

[3] CNRS, UMR 6602, IP, F-63171 Aubière, France

E-mail: vincent.rochatte@univ-bpclermont.fr

1. Introduction

Microalgae cultivation in photobioreactor is identified as a promising renewable way to produce a wide range of molecules for chemistry and energy. Nervertheless the industrial-scale utilisation of photobioreactors first requires the development of new technologies, or the optimisation of existing concepts, in order to reach thermodynamic efficiencies approaching at least 10%. This goal can only be achieved if robust models of the process are available for their design, sizing, optimization and operation, which require to address radiative transfer within the non-gray absorbing and anisotropically scattering microalgae suspension.

Furthemore it is nowadays well established that state-of-the-art photobioreactors are governed by radiative transfer within the visible and near infrared spectrum (photosynthetically active radiation), determining the growth rate, kinetic yield and biomass composition [1][2]. Therefore optimisation of the productivity and efficiency requires to control the radiation field within the reaction volume, which can be achieved by an appropriated design focused on the photobioreactor's geometry and boundary conditions (incident flux and materials' reflectivities). Here such a model-based optimisation is illustrated by the study of a cylindrical reactor prototype in which the incident light flux density is diluted in the culture volume thanks to a thousand of light-diffusing optical fibres emitting a quasi homogeneous density flux on the totality of their lateral immersed surface (see figure 1).

The present work addresses the implementation of a model of photobioreactors that is synthesised in figure 2. This implementation aims at simulating any photobioreactor's geometry in order to test its performances. Multiple scattering radiative transfer in complex geometry is solved in section 2 thanks to a Monte Carlo algorithm presented in [3] that is implemented here in the Computer Aided Design geometry of the DiCoFluV reactor (see figure 1). This resolution of the radiative transfer equation and its coupling with kinetics of photosynthesis is used in section 3 to simulate a culture of the photosynthetic microorganism *Arthrospira platensis*. Comparison between simulated results and experimental measurements will be presented and discussed during the seminar.

2. Implementation

Here we describe the implementation in complex geometry of the radiative transfer Monte Carlo algorithm presented in [3]. This algorithm is based on the most recent advances in the field of radiative transfer Monte Carlo [3][5][6], including integral formulation, sensitivity analyses and zero-variance approaches. It rigorously estimates the amount of photons absorbed locally per unit time per microalgae (integrated over the Photosynthetically Active Radiation spectral-domain) at any location **x** within the culture volume. The algorithm consists in the backward sampling of multiple-scattering and reflection optical-paths from **x** (the absorption location) to the emitting surface (the fibres surface: see figure 1). This algorithm is implemented in the EDStar development environment [5][7], that makes available to radiation physicists a set of computation tools issued from the computer graphics research community during the last 20 years, in particular in the framework of the Physically Based Rendering Techniques (PBRT) project [8]. Abstractions such as *ray, intersection, shape, etc.* are available to track photons in a given geometry considering absorption, scattering, reflection, etc.

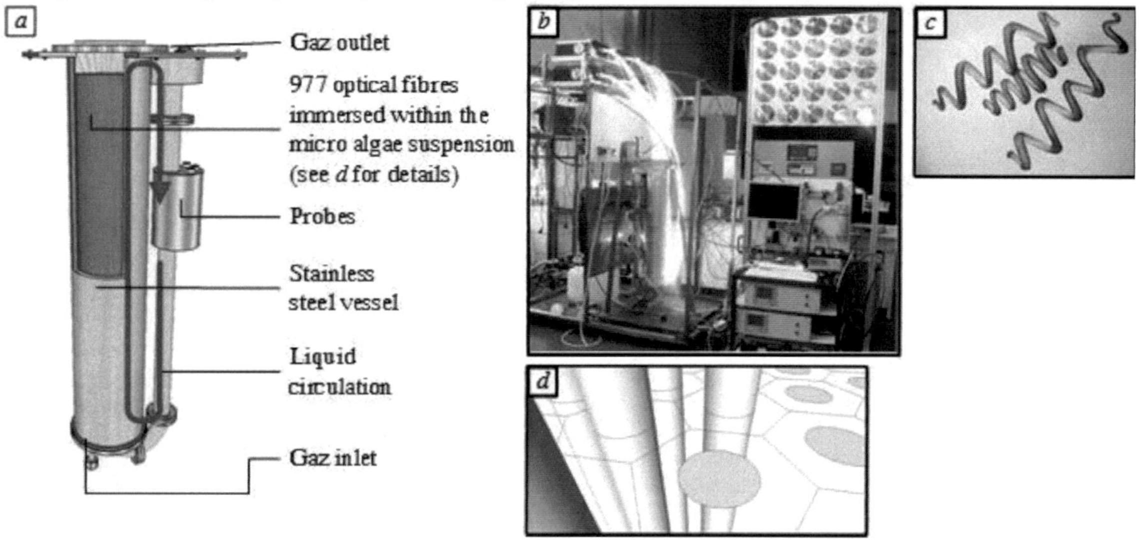

Figure 1 : DiCoFluV photobioreactor [4] in which incident light provided by discharge lamps is diluted within a 21L *Arthrospira platensis* culture (see micrograph c) via 977 lateral diffusing optical fibres (with reflectivity 0.1) emitting uniform surface flux density within the visible spectrum only. The stainless steel vessel has reflectivity 0.5. No emission within the volume is considered: fluorescence is negligible and only visible wavelengths are addressed. The liquid phase is homogenized with an airlift with a gas volumetric flow rate set at 21 L/min (1 vvm): the culture volume is homogeneous with absorption and scattering properties reported in [5]. (a) Computer Aided Design. (b) The complete pilot plant. (d) Close up on the hexagonal fibres lattice.

Here we explore our ability to extend this practice to the simulation of the performances of a reactor based on its Computer Aided Design (CAD) that is the tool used by engineers in the process of designing new innovative reactors, from the initial sketching to the final realisation. We imported the CAD of the DiCoFluV prototype within EDStar and estimated the radiation field within this three dimensional geometry that is composed of 73 000 triangles describing the surfaces of fibres and stainless steel vessel (see figure 1). Common bugs encountered in such complex geometry were tracked, in particular photon losses at the triangles' edges. No major difficulties were encountered in the present case. Moreover we recorded stable calculation-times when increasing geometric complexity.

3. Biomass productivity and thermodynamic efficiency modelling

Arthrospira platensis cultures have already been put in operation in the DiCoFluV reactor. The aim is to measure experimentally the volumetric productivity (in $g.h^{-1}.m^{-3}$) of the reactor operating at steady state. Experimental relation between microalgae concentration and productivity will be investigated. An optimum is expected: it would correspond to a situation where all the photons emitted are absorbed by microalgae. A first steady state has already been maintained for seven days (micoralgae concentration: $4g.L^{-1}$) corresponding to a volumetric productivity of around 0.02 $g.L^{-1}.h^{-1}$ and an efficiency of 13% (calculated as described in [9]). Further experimental work will be presented at the congress. The simulation associated to those measurements will be presented: it will include multiple anisotropic scattering (see section 2) and the non linear kinetic coupling. The whole methodological chain (see figure 2) will be used from the pigments concentrations and size distribution measurements to the calculation of the productivity which will be compared to the experimental results.

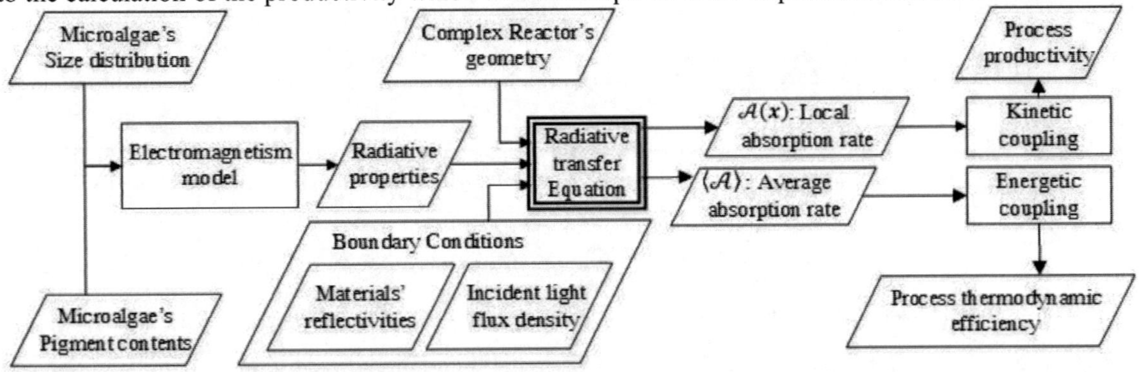

Figure 2 : The methodological chain. Rectangles: theories or models. Parallelograms: data. Central rectangle: radiative transfer Monte Carlo implementation described in section 2. Radiative properties (defining absorption and scattering properties of the participating media) are calculated, based on microalgae's pigment contents and size distribution thanks to an electromagnetic model that is out of the scope of the present article.

To conclude, a radiative transfer model that considers phenomena such as absorption and diffusion by microalgae and reflections on surfaces in photobioreactors is proposed. Thanks to the most recent advances for the implementation of Monte Carlo algorithms this model can be implemented in complex geometries for the evaluation of photobioreactors' performances. The model will be compared to experimental data.

Acknowledgements
This work has been sponsored by the French government research program Investissements d'avenir through the IMobS[3] Laboratory of Excellence (ANR-10-LABX-16-01), the ANR projects Biosolis (2008-11) and PRIAM (2013-15), by the European Union through the programm Regional competitiveness and employment 2007-2013 (ERDF – Auvergne region), and by the Auvergne region. We acknowledge Méso-Star SAS for fruitful discussion about ray tracing in complex geometry.

References

[1] Aiba S 1982 *Adv Biochem Eng.* **23** 85–156
Pilon L, Berberoğlu H, Kandilian R 2011 J *Quant. Spectrosc. Radiat. Transf.* **112** 2639–60
[2] Cornet J F 2010 *Chemical Engineering Science* **65** 985–998
[3] Dauchet J, Blanco S, Cornet J F, El-Hafi M, Eymet V and Fournier R 2013 *JQSRT* **128** 52–59
[4] Cornet J F, Lafon P, Joyard F, Goetz V, Plantard G, Garcia R 2012 International Patent N° WO/2012/069622
[5] Delatorre J *et al.* 2014 *Solar Energy* **103** 653–681
[6] Roger M, Blanco S, El Hafi M and Fournier R 2005 *Phys. Rev. Lett.* **95** 180601
Modest M F 2003 *Radiative Heat Transfer* (Academic Press, 2nd Edition)
[7] StarWest, EDStar development environement project, http://edstar.lmd.jussieu.fr/
[8] Pharr M and Humphreys G 2010 *Physically Based Rendering: From Theory to Implementation* (Morgan Kaufmann ed., 2nd Edition)
[9] Cornet J F, Dussap C G, Gros J B 1994 *AIChE* **40** 1055–66

The Generalized SLW Model

Vladimir P Solovjov[1], Frédéric Andre[2], Denis Lemonnier[3], Brent W Webb[1]

[1]Brigham Young University, CTB-435, Provo, UT 84602, USA
[2]Centre de Thermique et d'Energétique de Lyon, INSA de Lyon – 9 rue de la Physique – 69621 Villeurbanne, France
[3]ISAE-ENSMA, BP 40109, 86961 Futuroscope Chasseneuil Cedex, France

E-mail: vps@et.byu.edu

Abstract. The Generalized SLW Method is presented, formulating the SLW method with the help of both the ALBDF and the Inverse ALBDF. The result is two equivalent symmetric models: the SLW Model and the Inverse SLW Model. The advantage of the unified dual formulation and of application of the ALBDF and the Inverse ALBDF is in more efficient implementation of the model and the elimination of the solution of the implicit equations for the absorption cross-sections in the construction of the spectral model in the case of non-isothermal media. The generalized approach explores all possibilities of the SLW method under both direct and inverse formulations including its limiting cases: the minimal one clear gas–one gray gas SLW-1 model, and the case when the number of gray gases approaches infinity termed the Exact SLW model. The present work outlines the steps in a unified construction of the generalized SLW model in isothermal and non-isothermal media, and compares different forms of the modelled radiative quantities in plane parallel media: directional total radiative flux, total emissivity, Planck mean and Rosseland mean absorption coefficients.

1. Introduction

The Spectral Line Weighted-sum-of-gray-gases (SLW) method is the first global approach for spectral modeling of radiative transfer in high temperature gases developed on the basis of the high resolution gas absorption spectrum [1-21]. Since that time the method has undergone a number of extensions, interpretations and variations, and other global models have been introduced: ADF, FSK, CW and MBWSGG methods, and hybrid forms of the SLW method with other spectral models such as EWBM, reordered wavenumber, correlated *k*-distributions, cumulative wavenumber, etc. [22,31-33,34-37].

The primary basis in accounting for spectral quantities in the SLW method is the distribution function $F\left(C,T_g,T_b,Y,p\right)$ of a continuous cross-section variable C termed the Absorption Line Blackbody Distribution Function (ALBDF). The ALBDF is the characterization of the high resolution gas absorption cross-section $C_\eta\left(T_g\right)$ at gas temperature T_g, gas molar fraction Y, and total pressure p, regarding the Planck spectral distribution of the blackbody emissive power of radiation at the source temperature T_b. The present work is a unified formulation of the spectral line weighted-sum-of-gray-gases (SLW) approach to spectral modelling of gas radiation based on an alternative application of both the direct ALBDF and the inverse ALBDF. This yields two independent but completely equivalent symmetric gray gas spectral models which make the SLW method more universal, efficient, and

Journal of Physics: Conference Series **676** (2016) 012022 doi:10.1088/1742-6596/676/1/012022

flexible. Both spectral models have a one-to-one correspondence between discretization of the cross-section variable C (direct SLW model) and the ALBDF F (inverse SLW model). Simultaneous synchronized discretization of both variables allows reduction of the number of gray gases to 3 - 6 gray gases without significant loss of accuracy, avoiding costly optimization. This approach is named here the compact SLW model. The resulting reordered absorption cross-section of the SLW spectral model allows clear definition and interpretation of the total characterization of the gas absorption coefficient such as total emissivity, and the Planck and the Rosseland mean absorption coefficients.

In the case of non-uniform media the SLW reference approach, based on assumption of correlation between absorption spectra at different states, alternates between applications of four direct and inverse ALBDFs calculated at the reference and the local states, providing more choices for better implementation of the SLW method. The advantage of this unified modelling approach is the elimination of the solution of implicit equations traditionally used in the reference approach. The Generalized SLW Method also includes two limiting cases: the minimal SLW model, which consists of a single clear gas – single gray gas model; and the Exact SLW model, in which the number of gray gases approaches infinity. These limiting cases also have alternative direct and inverse presentations.

This paper presents a unified treatment of the SLW direct and inverse methods in a single location, drawing some from previously published work as review to set the foundation for new material, and developing new explorations for other scenarios. The paper includes considerable detail, intending to serve as a reference for those seeking to implement one of the family of SLW models.

2. Absorption Line Blackbody Distribution Function (ALBDF) and Inverse ALBDF

2.1 Radiative Transfer Equation, gas absorption coefficient and the gas absorption cross-section

The spectral radiation intensity $I_\eta(s,\mathbf{\Omega})$ in absorbing and emitting media along a pathlength s in a direction $\mathbf{\Omega}$ is described by the Radiative Transfer Equation (RTE):

$$\frac{\partial I_\eta(s,\mathbf{\Omega})}{\partial s} = -\kappa_\eta(s)I_\eta(s,\mathbf{\Omega})+\kappa_\eta(s)I_{b\eta}\left[T(s)\right]$$

where $I_{b\eta}\left[T(s)\right]$ is the Planck blackbody intensity at the local gas temperature $T(s)$, and $\kappa_\eta(s)$ is the local spectral absorption coefficient. The construction of the SLW spectral model is typically based on the gas absorption cross-section $C_\eta(T_g,Y)$ rather than on the gas absorption coefficient. Consider the dependence of the gas absorption coefficient on its thermodynamic state (total pressure p, gas partial pressure p_g, molar fraction Y, gas temperature T_g, molar density N):

$$\kappa_\eta = \underbrace{\left(\frac{p}{R_uT_g}\right)\left[\frac{mol}{m^3}\right]}_{N}\cdot\underbrace{\left(\frac{p_g}{p}\right)}_{Y}\cdot\underbrace{N_A\left[\frac{molecule}{mol}\right]\cdot C_\eta(T_g,Y,p)\left[\frac{cm^2}{molecule}\right]\cdot\frac{1}{10{,}000}\left[\frac{m^2}{cm^2}\right]}_{C_\eta(T_g,Y,p)\left[\frac{m^2}{mol}\right]\ \text{spectral absorption cross-section}},\ \left[\frac{1}{m}\right]$$

The gas absorption cross-section $C_\eta(T_g,Y,p)$, $\left[m^2/mol\right]$ defined by this relationship is used in the SLW method for characterization of the gas absorption spectrum.

The Line-by-Line (LBL) method solves the spectral RTE for each wavenumber η, after which the following spectral integration for obtaining the total intensity of radiation is used:

$$I(s,\mathbf{\Omega})=\int_0^\infty I_\eta(s,\mathbf{\Omega})d\eta$$

The SLW method is a particular case of the class of so-called global methods in gas radiation which, in contrast to the Line-by-Line method, starts first with the spectral integration of the RTE, and then solves it for the spectrally integrated intensity.

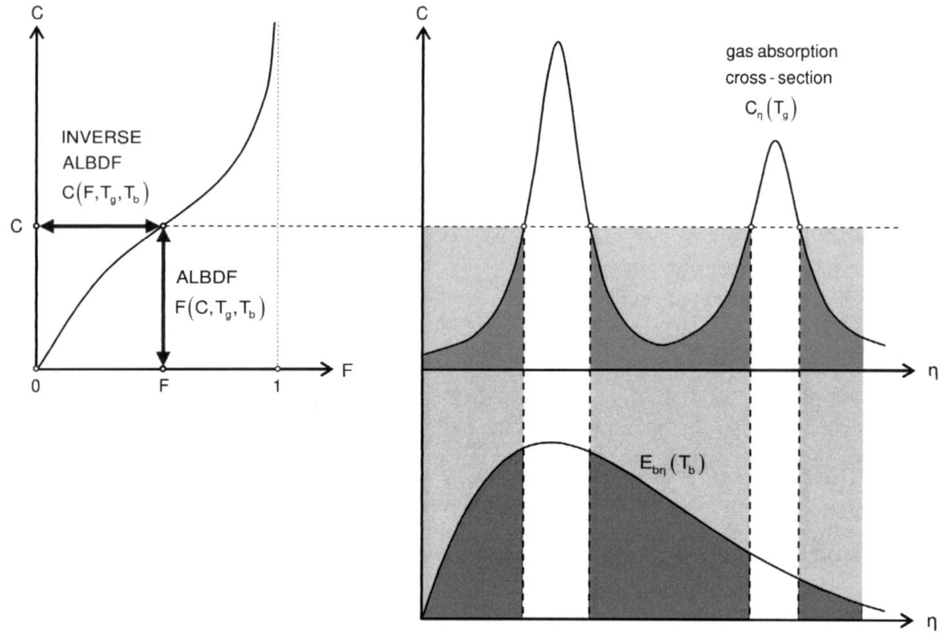

Figure 1. Definition of the direct ALBDF $F\left(C,T_g,T_b\right)$ and the inverse ALBDF $C\left(F,T_g,T_b\right)$

2.2. *Definition of the ALBDF and the Inverse ALBDF*

ALBDF $F\left(C,T_g,T_b,p,Y\right)$ is the distribution function which characterizes the high resolution line gas absorption spectrum $C_\eta\left(T_g,Y,p\right)$, $\left[m^2/mol\right]$ at a given thermodynamic state: gas temperature T_g, gas total pressure p, gas molar fraction Y. The function describes the fraction of the blackbody radiation power $E_b\left(T_b\right)=\sigma T_b^4$ emitted at temperature T_b that is in the part of the spectrum where the gas line absorption cross-section $C_\eta\left(T_g\right)$ is below the prescribed value C (see figure 1):

$$F\left(C,T_g,T_b,p,Y\right) = \frac{1}{E_b\left(T_b\right)} \int_{\left\{\eta:C_\eta\left(T_g,p,Y\right)<C\right\}} E_{b\eta}\left(T_b\right)d\eta$$

The ALBDF is an increasing monotonic function varying between 0 and 1 in variable *C*, and therefore, it is invertible.

Inverse ALBDF The inverse ALBDF $C\left(F,T_g,T_b\right)$ of the variable *F* is defined such that

$$F\left[C\left(F,T_g,T_b,p,Y\right),T_g,T_b,p,Y\right]=F \text{, and } C\left[F\left(C,T_g,T_b,p,Y\right),T_g,T_b,p,Y\right]=C$$

The inverse ALBDF defines the value of the absorption cross-section *C*, for which the prescribed value *F* is the fraction of the blackbody emissive power emitted at temperature T_b in the part of the spectrum where the absorption cross-section at the prescribed gas thermodynamic state is below *C*.

2.3. Tabulated LBL ALBDF

A finely resolved ALBDF look-up table has been developed previously [21] for H_2O, CO_2, and CO for a range of gas and blackbody source temperatures from 300K to 3000K (with increment 100K), and total pressures from 0.1 atm to 50 atm (at different mole fraction of H_2O for taking into account self and air-broadening). Calculations of the ALBDF were performed with the help of the HITEMP-2010 spectral database. An efficient algorithm has been used for fast multilinear interpolation and inversion for calculation of the direct and inverse ALBDF at the prescribed thermodynamic state. The accuracy of prediction based on the look-up table is nearly identical to LBL calculations, and can be used as a benchmark for verification of spectral models. The ALBDF look-up table is available for download at http://albdf.byu.edu .

2.4 ALBDF Correlations

Because of simplicity and convenience in their applications, the most popular representations of the ALBDF are mathematical correlations of the form developed by Denison and Webb [1,3,6]. Despite the now-obsolete spectral database on which earlier correlations were based, these correlations are quite accurate compared to updated and more complete spectral data [27]. Coefficients of correlations are presented in [1,3].

H_2O

$$F_w\left(C,T_g,T_b,Y_w\right) = \frac{1}{2}\tanh\left[P_w\left(T_g,T_b,\xi-\xi_{sb}\right)\right]+\frac{1}{2}$$

$$P_w\left(T_g,T_b,\xi-\xi_{sb}\right) = \sum_{l=0}^{3}\sum_{m=0}^{3}\sum_{n=0}^{3} b_{lmn}\left(\frac{T_g}{2500}\right)^n\left(\frac{T_b}{2500}\right)^m\left(\xi-\xi_{sb}\right)^l$$

$$\xi_{sb} = \sum_{l=0}^{3}\sum_{m=0}^{3}\sum_{n=0}^{2} c_{lmn}\left(\frac{T_b}{2500}\right)^n \xi^m\left(Y_w\right)^{l+1}$$

$$\xi=\ln(C), \qquad 1\cdot10^{-4}\le C\le 60\left[\frac{m^2}{mol}\right]$$

CO_2

$$F_c\left(C,T_g,T_b\right) = \frac{1}{2}\tanh\left[P_c\left(T_g,T_b,\xi\right)\right]+\frac{1}{2}$$

$$P_c\left(T_g,T_b,\xi\right) = \sum_{l=0}^{3}\sum_{m=0}^{3}\sum_{n=0}^{3} d_{lmn}\left(\frac{T_g}{2500}\right)^n\left(\frac{T_b}{2500}\right)^m\left(\xi\right)^l$$

$$\xi=\ln(C), \qquad 1\cdot10^{-4}\le C\le 600\left[\frac{m^2}{mol}\right]$$

In Pearson et al., [19], the HITEMP-2010 database was used to update the correlation coefficients b_{lmn} , c_{lmn}, d_{lmn}. Further, additional coefficients a_{lmn} for correlations of the carbon monoxide ALBDF were reported:

CO

$$F_m\left(C,T_g,T_b\right) = \frac{1}{2}\tanh\left[P_m\left(T_g,T_b,\xi\right)\right]+\frac{1}{2}$$

$$P_m\left(T_g,T_b,\xi\right) = \sum_{l=0}^{3}\sum_{m=0}^{3}\sum_{n=0}^{3} a_{lmn}\left(\frac{T_g}{2500}\right)^n\left(\frac{T_b}{2500}\right)^m\left(\xi\right)^l$$

$$\xi=\ln(C), \qquad 1\cdot10^{-4}\le C\le 60\left[\frac{m^2}{mol}\right]$$

Note that the correlation for H_2O includes the gas molar fraction Y_w to account for the influence of self-broadening on the ALBDF (which is handled by a shift ξ_{sb} of the variable ξ), whereas the ALBDF of CO_2 and of CO does not depend on gas molar fraction. The correlations of Denison and Webb [1,3,6] were developed with the help of the originally generated spectral database and they are valid for the range of temperatures $400K \le T_g, T_b \le 2500K$. Extrapolation of these correlations to temperatures as low as room temperature has been found to be sufficiently accurate for engineering applications. The updated correlations of Pearson et al. [19] were developed with the help of the updated HITEMP-2010 spectral database, and they are valid for a wider range of temperatures: $400K \le T_g, T_b \le 3000K$. In their most fundamental form, both sets of correlations are designated for application at standard atmospheric total pressure $p = 1\ atm$.

Correlations for the ALBDF at variable total pressure in the range $0.1\ atm \le p \le 50\ atm$ were developed by Pearson et al. [21]:

H_2O

$$F_w(C, T_g, T_b, Y_w, p) = \frac{1}{2}\tanh\left[P_w(T_g, T_b, \xi - \xi_p)\right] + \frac{1}{2}$$

$$P_w(T_g, T_b, \xi - \xi_p) = \sum_{l=0}^{3}\sum_{m=0}^{3}\sum_{n=0}^{3} b_{lmn}\left(\frac{T_g}{2500}\right)^n\left(\frac{T_b}{2500}\right)^m(\xi - \xi_p)^l$$

$$\xi_p = \sum_{l=0}^{3}\sum_{m=0}^{3}\sum_{n=0}^{3} u_{lmn}\left(\frac{T_g T_b}{2500^2}\right)^n \xi^m \psi_w^{l+1}$$

$$\psi_w = \frac{1}{10}\ln(100 p_e)$$

$$p_e = (1 + 8.17 \cdot Y_w)p$$

$$\xi = \ln(C), \qquad 1 \cdot 10^{-4} \le C \le 60 \left[\frac{m^2}{mol}\right]$$

CO_2

$$F_c(C, T_g, T_b, p) = \frac{1}{2}\tanh\left[P_c(T_g, T_b, \xi - \xi_p)\right] + \frac{1}{2}$$

$$P_c(T_g, T_b, \xi) = \sum_{l=0}^{3}\sum_{m=0}^{3}\sum_{n=0}^{3} d_{lmn}\left(\frac{T_g}{2500}\right)^n\left(\frac{T_b}{2500}\right)^m(\xi - \xi_p)^l$$

$$\xi_p = \sum_{l=0}^{3}\sum_{m=0}^{3}\sum_{n=0}^{3} v_{lmn}\left(\frac{T_g T_b}{2500^2}\right)^n \xi^m \psi_c^{l+1}$$

$$\psi_c = \frac{1}{10}\ln(100 p)$$

$$\xi = \ln(C), \qquad 1 \cdot 10^{-4} \le C \le 600 \left[\frac{m^2}{mol}\right]$$

CO

$$F_m\left(C,T_g,T_b,p\right) = \frac{1}{2}\tanh\left[P_m\left(T_g,T_b,\xi\right)-P_p\left(T_g,T_b,\xi\right)\right]+\frac{1}{2}$$

$$P_m\left(T_g,T_b,\xi\right) = \sum_{l=0}^{3}\sum_{m=0}^{3}\sum_{n=0}^{3} a_{lmn}\left(\frac{T_g}{2500}\right)^n\left(\frac{T_b}{2500}\right)^m\left(\xi\right)^l$$

$$P_p\left(T_g,T_b,\xi\right) = \sum_{l=0}^{3}\sum_{m=0}^{3}\sum_{n=0}^{3} w_{lmn}\left(\frac{T_gT_b}{2500^2}\right)^n\xi^m\psi^{l+1}$$

$$\psi = \frac{1}{10}\ln\left(100p\right)$$

$$\xi = \ln\left(C\right), \qquad 1\cdot 10^{-4} \le C \le 60 \left[\frac{m^2}{mol}\right]$$

The correlation coefficients at atmospheric pressure b_{lmn}, c_{lmn}, d_{lmn}, a_{lmn}, and the coefficients for the correlations at variable total pressure u_{lmn}, v_{lmn}, and w_{lmn} are available for download at http://albdf.byu.edu , together with the C++ code for calculation of the ALBDF using the correlations. The look-up table for the ALBDF generated from LBL spectral data used for development of correlations can also be found at this website. Other correlations for the ALBDF are developed in [22,26,27,29].

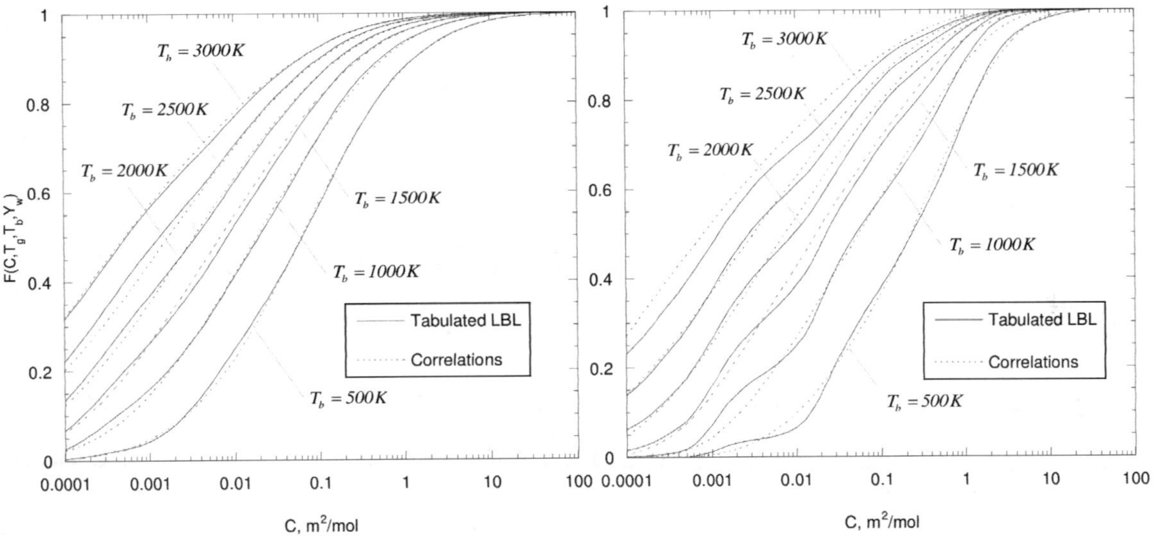

Figure 2. Example of comparison of correlations and look-up table data [21]**.** Y_w=0 prescribes that in calculation of absorption coefficient only air broadening of spectral lines is taken into account (no self-broadening)*:* a) H_2O (Y_w=0), T_g=1000K, p=1atm; b) H_2O (Yw=0), T_g=1000K, p=20atm

As seen in figure 2, the mathematical correlations of the ALBDF described previously reproduce the behavior of the ALBDF curves quite well and for most conditions. However, they deviate from the LBL data under some conditions. Note that dependence of the ALBDF on cross-section C is shown in logarithmic scale with six orders of magnitude variation. The steep growth of ALBDF with C prevents application of traditional approximations, for example, with the help of orthogonal polynomials which yield very strong oscillations of approximations. The problem of accurate polynomial representation of

ALBDF may be effectively solved with the help of the Method of Moments and Cutteridge-Devyatov Polynomials (CDP) in logarithmic scale.

2.5 Interpolation and inversion of the tabulated LBL ALBDF with the help of Method of Moments

The following approach provides the ALBDF as a continuous function of absorption cross-section C utilizing the Method of Moments for polynomial representation of the functions. The basics of the method are developed and presented in [24,44], where solution of the Classical Moment Problem on a finite interval is formulated in the following form:

The Cutteridge-Devyatov polynomials (CDP) of order n, $\Phi_k^{(n)}(x)$, are defined as

$$\Phi_k^{(n)}(x) = \sum_{m=0}^{n-1}(-1)^{k+m}\cdot\frac{(n+k)!}{k!(k+1)!(n-k-1)!}\cdot\frac{(n+m)!}{m!(m+1)!(n-m-1)!}\cdot\frac{k+1}{k+m+1}\cdot x^{m+1}, \quad k=0,1,...,n-1$$

Let $f(x)$, $x\in[0,1]$ be a function such that $f(0)=0$, and let the first n generalized moments of $f(x)$ be defined by

$$J_k=\int_0^1 x^k d\left[f(x)\right], \quad k=0,1,...,n-1$$

Let the function $f_n(x)$ be defined as a linear combination of n polynomials $\Phi_k^{(n)}(x)$, $k=0,1,...,n-1$:

$$f_n(x) = \sum_{k=0}^{n-1} J_k\Phi_k^{(n)}(x),$$

where the coefficients J_k are the generalized moments of $f(x)$. Then the first n generalized moments of $f_n(x)$ and $f(x)$ coincide, and the sequence $f_n(x)\to f(x)$ approaches the continuous function. The outline of representation of the ALBDF with the help of CDP is the following:

One uses multilinear interpolation to read ALBDF data from the look-up table at a given thermodynamic state $\{T_g,T_b,p,Y\}$: $F_j=F(C_j,T_g,T_b,p,Y)$, $j=0,1,2,...,N$, where N is the number of points in the table. Denote $C_{\min}=C_0$, $C_{max}=C_N$ and $F_{min}=F_0$, $F_{max}=F_N$. Assume that $F(C,T_g,T_b,p,Y)$ is a linear function of the variable C in the intervals $\left[C_{j-1},C_j\right]$, $j=1,...,N$. The function $F(C,T_g,T_b,p,Y)$, by its definition, is strictly increasing on the entire interval $\left[C_{\min},C_{\max}\right]$, and therefore, it is invertible.

Logarithmic change of variable and normalization is performed as follows:

$$c_j=\ln\frac{C_j}{C_{\min}}\Big/\ln\frac{C_{\max}}{C_{\min}}, \qquad f_j=\frac{F_j-F_{\min}}{F_{\max}-F_{\min}}, \qquad j=0,1,2,...,N$$

Assume the function $f(c)$, $c\in[0,1]$, $f(0)=0$ and $f(1)=1$ to be piecewise linear in the intervals $\left[C_{j-1},C_j\right]$, $j=1,...,N$:

$$f(c) = \sum_{j=1}^{N}\left[\frac{f_j-f_{j-1}}{c_j-c_{j-1}}(c-c_{j-1})+f_{j-1}\right]\cdot\left[H(c-c_{j-1})-H(c-c_j)\right], \; c\in[0,1]$$

where $H(c)$ is the Heaviside unit step-function. The normalized function $f(c)$ is strictly increasing, and therefore, it is invertible. The inverse function is denoted as $c(f)$. An example of both the normalized direct and inverse functions for CO_2 at $T_g = T_b = 1000K$ are shown in figure 3.

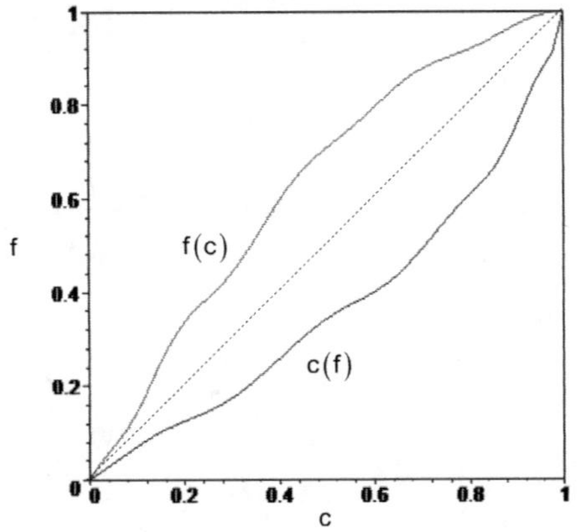

Figure 3. The normalized direct *f(c)* and inverse *c(f)* functions

One may calculate the generalized moments of $f(c)$ and of the inverse function $c(f)$ accounting for the fact that both functions are piecewise linear:

$$J_k = \int_0^1 c^k d\left[f(c)\right] = \frac{1}{k+1}\cdot\sum_{j=1}^{N}\frac{c_j^{k+1}-c_{j-1}^{k+1}}{c_j - c_{j-1}}\cdot\left(f_j - f_{j-1}\right), \qquad k = 0,1,\ldots,n-1$$

$$M_k = \int_0^1 f^k d\left[c(f)\right] = \frac{1}{k+1}\cdot\sum_{j=1}^{N}\frac{f_j^{k+1}-f_{j-1}^{k+1}}{f_j - f_{j-1}}\cdot\left(c_j - c_{j-1}\right), \qquad k = 0,1,\ldots,n-1$$

Further, the function $f(c)$ and the inverse function $c(f)$ may be represented with the help of Cutteridge-Devyatov Polynomials:

$$f(c) = \sum_{k=0}^{n-1} J_k \Phi_k^{(n)}(c)$$

$$c(f) = \sum_{k=0}^{n-1} M_k \Phi_k^{(n)}(f)$$

Changing the variables c and f back to C and F to represent the ALBDF and the inverse ALBDF yields:

$$F\left(C,T_g,T_b,p,Y\right) = F_{\min} + \left(F_{\max} - F_{\min}\right)\sum_{k=0}^{n-1} J_k \Phi_k^{(n)}\left(\ln\frac{C}{C_{\min}} \Big/ \ln\frac{C_{\max}}{C_{\min}}\right)$$

$$C\left(F,T_g,T_b,p,Y\right) = C_{\min}\cdot\left(\frac{C_{\max}}{C_{\min}}\right)^{\sum_{k=0}^{n-1} M_k \Phi_k^{(n)}\left(\frac{F - F_{\min}}{F_{\max} - F_{\min}}\right)}$$

With a sufficient number of moments (usually, 6 - 14), CDP provides very accurate and efficient representation of the ALBDF as shown in figures 4 and 5.

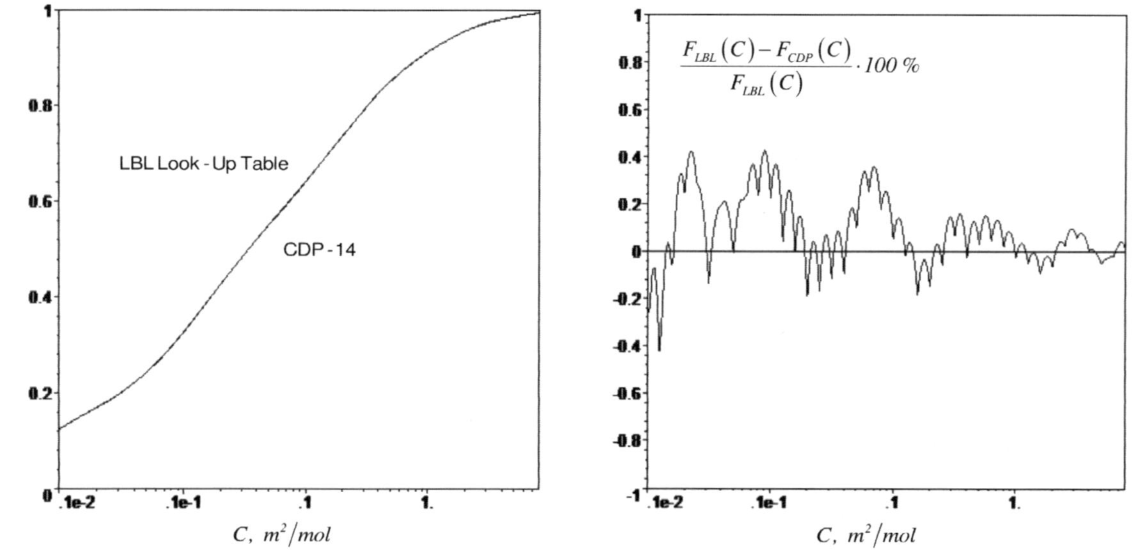

Figure 4. a) Comparison of the ALBDF look-up table data for H_2O at $T_g = T_b = 1000\ K$ with the CDP representation with the help of 14 moments; b) relative error.

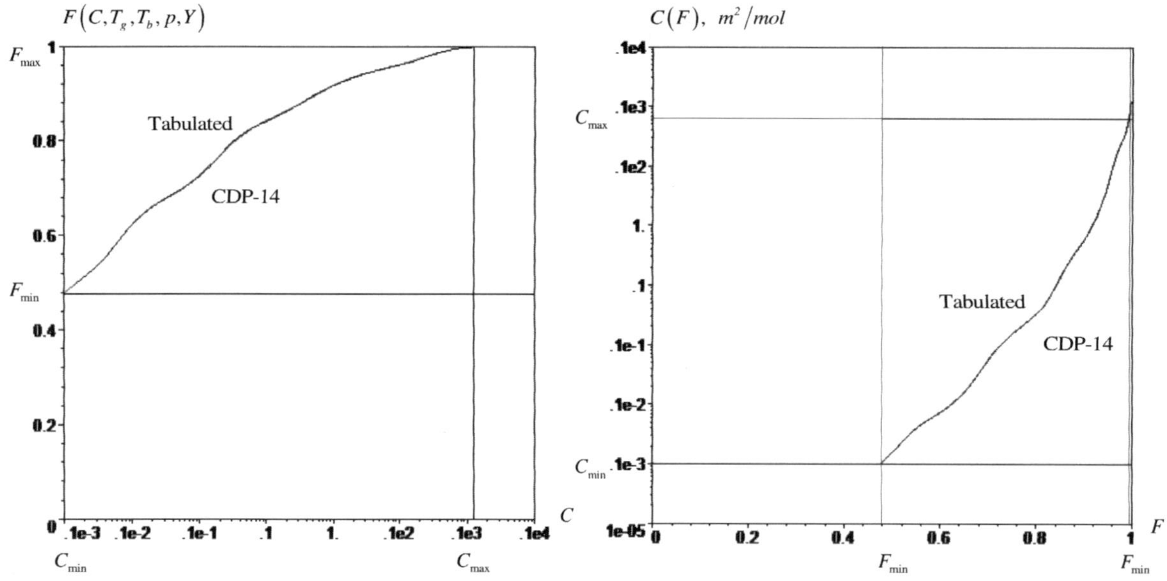

Figure 5. Comparison of the ALBDF look-up table data (a) and the inverse ALBDF (b) of CO_2 at $T_g = T_b = 1000\ K$ with the CDP representation.

3. Construction of the Generalized SLW Spectral Model

3.1 Simultaneous subdivision of the variables C and F

The generalized SLW spectral model is based on simultaneous synchronized subdivision of the variable C into supplemental absorption cross sections $\tilde{C}_0, \tilde{C}_1, ..., \tilde{C}_n$ and the subdivision of the variable F into corresponding values of $\tilde{F}_0, \tilde{F}_1, ..., \tilde{F}_n$. The supplemental cross-sections define the wavenumber intervals of gray gases for spectral integration. That subdivision is followed by the corresponding subdivision into absorption cross-sections $C_1, ..., C_n$ and associated values of $F_1, ..., F_n$. The absorption cross-sections define the values of the gray gas absorption coefficients $\kappa_j = NYC_j$.

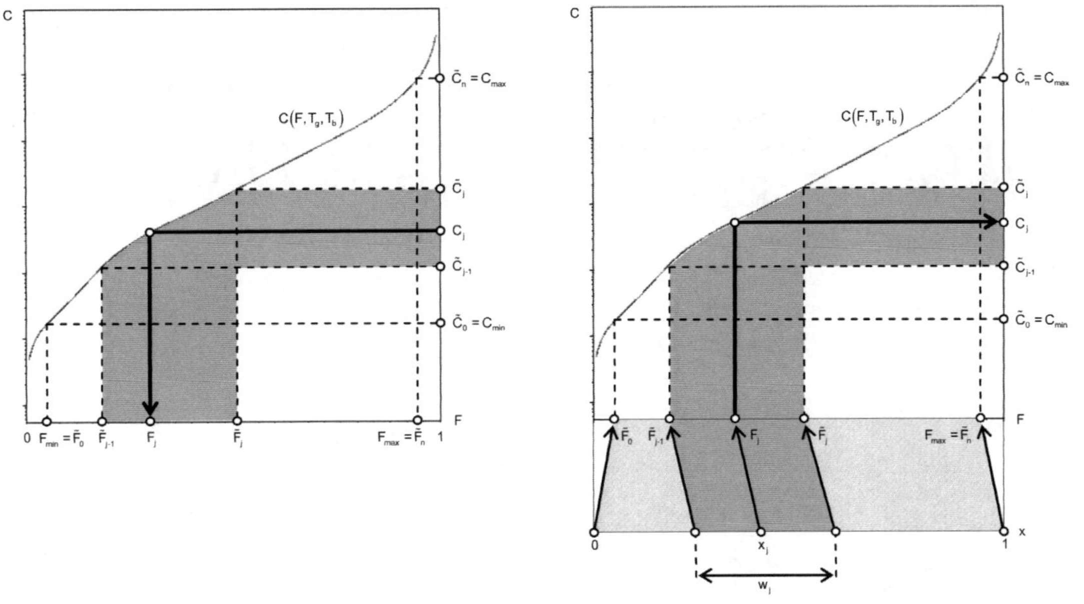

Figure 6. Simultaneous partition of the variables C and F: a) log-spaced absorption cross-sections $\tilde{C}_j$ generate partition of $\tilde{F}_j$; b) partition $\tilde{F}_j$ based on Gauss-Legendre quadratures generates partition $\tilde{C}_j$.

Construction of the standard SLW spectral model starts with subdivision of variable C into evenly log-spaced supplemental absorption cross sections $\tilde{C}_0, \tilde{C}_1, ..., \tilde{C}_n$, as shown schematically in Fig. 6. Generally, one chooses the range C_{min}, C_{max} which effectively represents the absorption cross-section of gases. Then the supplemental cross-sections are defined by

$$\tilde{C}_j = C_{min}\left(C_{max}/C_{min}\right)^{j/n}, \qquad j = 0, 1, ..., n$$

The absorption cross-sections for representation of the absorption coefficient as the geometrical mean of corresponding supplemental cross-sections are shown as:

$$C_j = \sqrt{\tilde{C}_{j-1}\tilde{C}_j} \qquad j = 1, ..., n$$

The gray gas absorption coefficients are then calcualtred as

$$\kappa_j = NYC_j \qquad j = 1,...,n$$

and the clear gas absorption coefficient is $\kappa_0 = 0$. Partition of the absorption cross-section variable C automatically generates the corresponding subdivision of variable F:

$$\tilde{F}_j = F\left(\tilde{C}_j, T_g, T_b\right) \qquad j = 0,1,...,n$$

$$F_j = F\left(C_j, T_g, T_b\right) \qquad j = 1,...,n$$

3.2 Outline of the compact SLW-Gauss model – construction of the SLW model in isothermal media

This approach is based on discretization of the ALBDF domain of the variable $F \in \left[F_{\min}, F_{\max}\right]$ with the help of Gauss-Legendre quadratures, as shown in the right panel of figure 6. This approach is called compact, because, in general, without optimization, for satisfactory accuracy it requires fewer gray gases (3-5) than the traditional logarithmically spaced subdivision in the SLW method (8-10 gray gases). The procedure is summarized as follows:

1) Choose n = the number of gray gases in the model

2) Define $\tilde{C}_0 = C_{min}$, $\tilde{C}_n = C_{max}$

3) Calculate
$$F_{min} = \tilde{F}_0 = F\left(C_{min}, T_g = T, T_b = T\right)$$
$$F_{max} = \tilde{F}_n = F\left(C_{max}, T_g = T, T_b = T\right)$$

4) Find positive abscissas $x_j > 0$ and weights w_j, $j = 1,2,...,n$ of the Gaussian-Legendre quadratures for integration over the interval $[-1, 1]$.

5) Discretize the interval $\left[F_{\min}, F_{\max}\right]$ by transformation of the Gaussian quadratures:
$$\tilde{F}_0 = F_{min}$$
$$\tilde{F}_j = F_{min} + \left(\sum_{k=1}^{j} w_k\right) \cdot \left(F_{max} - F_{min}\right) \qquad j = 1,2,...,n$$
$$F_j = F_{min} + x_j \cdot \left(F_{max} - F_{min}\right) \qquad j = 1,2,...,n$$

6) Calculate the supplemental absorption cross-sections with the help of the inverse ALBDF:
$$\tilde{C}_j = C\left(\tilde{F}_j, T_g = T, T_b = T\right) \qquad j = 1,2,...,n$$

7) Calculate the gray gas absorption cross-sections with the help of the inverse ALBDF:
$$C_j = C\left(F_j, T_g = T, T_b = T\right) \qquad j = 1,2,...,n$$

8) Calculate the gray gas absorption coefficients:
$$\kappa_0 = 0, \quad \kappa_j = NYC_j \qquad j = 1,2,...,n$$

9) Calculate the gray gas weights:
$$a_0 = \tilde{F}_0, \quad a_j = \tilde{F}_j - \tilde{F}_{j-1} \qquad j = 1,2,...,n$$

10) Calculate the boundary gray gas weights:

$$a_j^w = F\left(\tilde{C}_j, T_g = T, T_b = T_w\right) - F\left(\tilde{C}_{j-1}, T_g = T, T_b = T_w\right)$$

$$a_0^w = F\left(\tilde{C}_0, T_g = T, T_b = T_w\right)$$

11) The clear gas and the gray gas RTEs:

$$\frac{\partial I_0(s,\mathbf{\Omega})}{\partial s} = 0$$

$$\frac{\partial I_j(s,\mathbf{\Omega})}{\partial s} = -\kappa_j I_j(s,\mathbf{\Omega}) + a_j \kappa_j I_b(T)$$

subject to the corresponding boundary conditions.

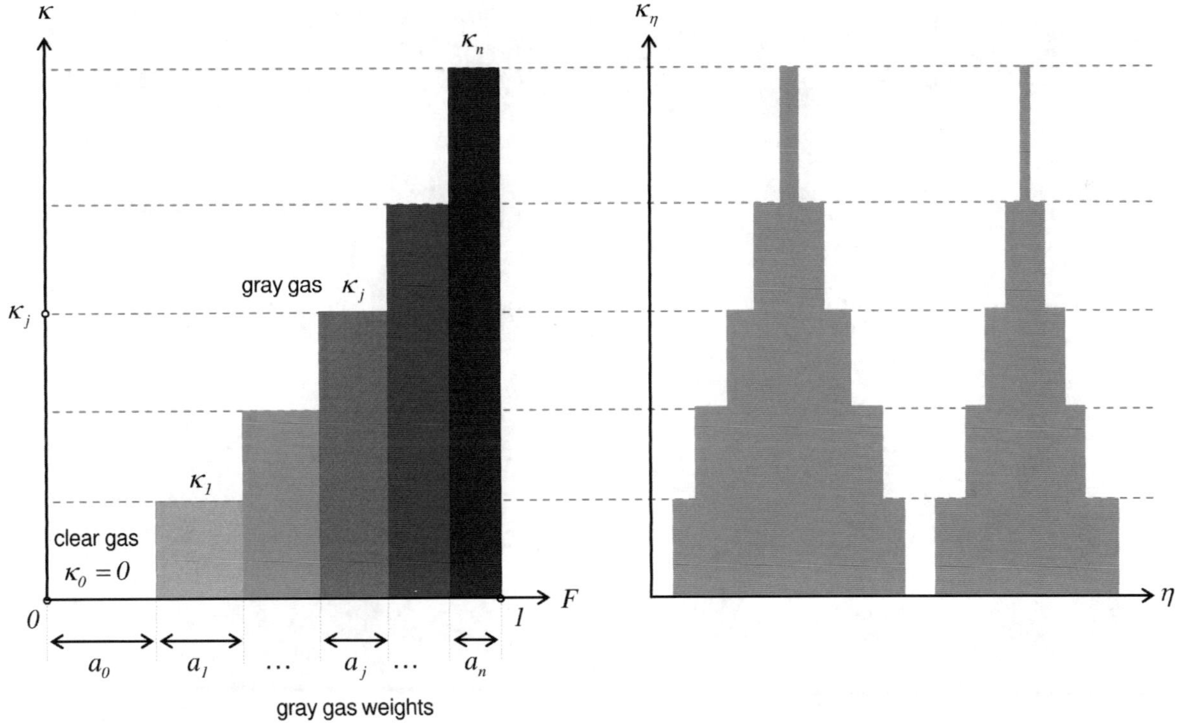

Figure 7. The SLW reordered gray gas absorption coefficients and the SLW histogram spectrum.

The SLW method turns the continuous line gas absorption spectrum into a histogram spectrum with a few dicrete values of the gray gas absorption coefficient $\kappa_j = NYC_j$, as illustrated in the right panel of figure 7. Then these gray gas absorption coefficients are reordered according to their weights a_j (left panel of figure 7). With increase of the number of gray gases, the reordered distribution of κ_j is approaching multiplied by gas molar density and fraction the inverse ALBDF $NYC\left(F, T_g, T_b\right)$ defined in the interval $F \in \left[F_{min}, F_{max}\right]$.

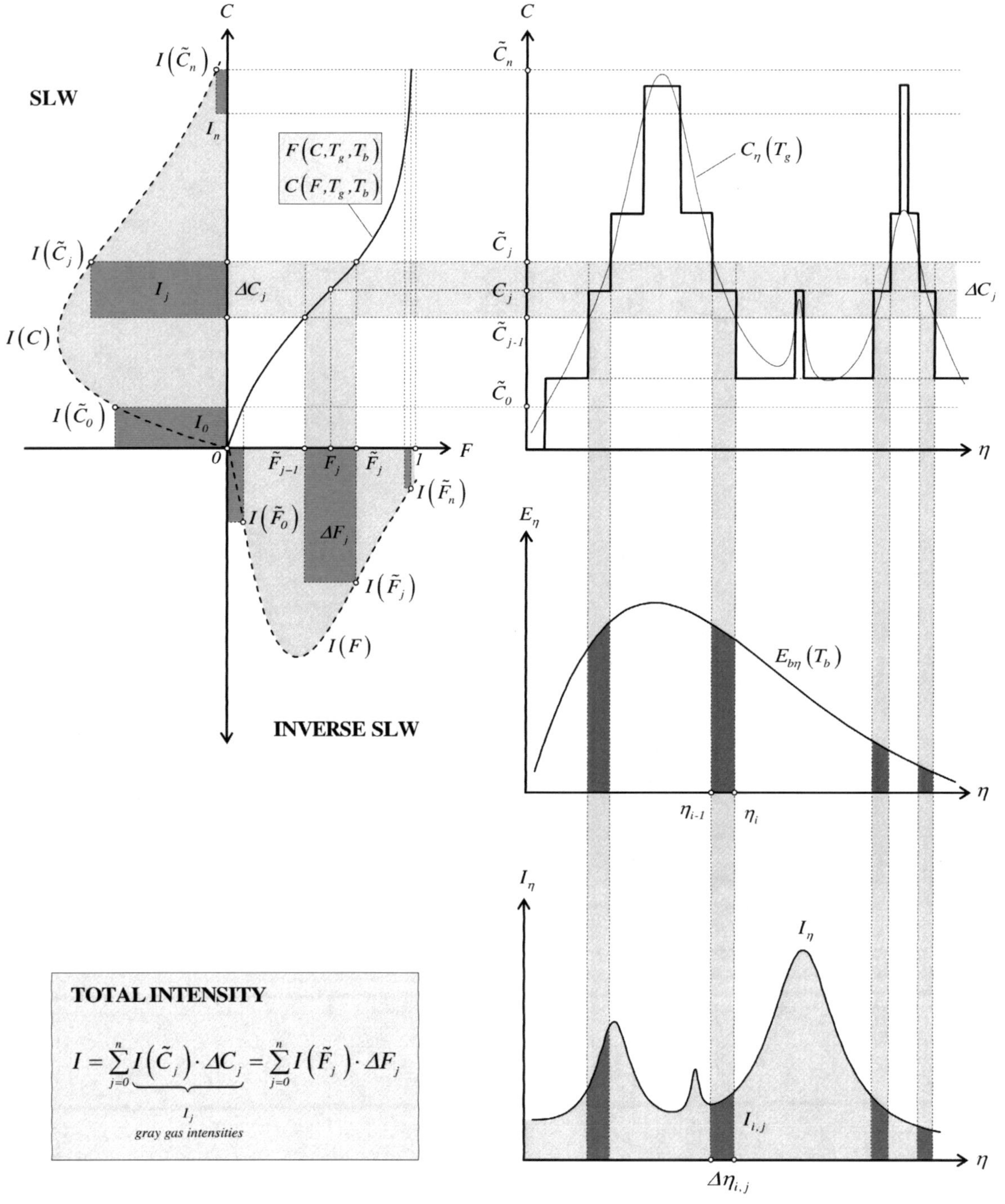

Figure 8. Geometric Interpretation of the SLW Method and of the Inverse SLW Method

3.3 Summary of the SLW spectral model in isothermal medium

The SLW and the inverse SLW spectral models based on simultaneous discretisation of variables C and F are visualized in figure 8. There is a clear one-to-one correspondence between two models. Table 1 compiles all possible forms of the gray gas RTE in the generalized formulation with the help of direct and inverse ALBDF. The analytical solution for the directional radiative flux $q^+(x)$ in a plane parrallel layer bounded by cold, black walls illustrates the differences and similarities in the different variations of the SLW method.

Table 1. SLW spectral models in isothermal media

Model	*RTE*	*Total Intensity* $I(s,\mathbf{\Omega})$	*Directional Radiative Flux* $q^+(x)$
LBL	$\frac{\partial I_\eta}{\partial s} = -\kappa_\eta I_\eta + \kappa_\eta I_{b\eta}(T)$	$\int_0^\infty I_\eta d\eta$	$\pi I_b - 2\pi \int_0^\infty I_{b\eta}(T) E_3(NYC_\eta x) d\eta$
SLW	$\frac{\partial I_j}{\partial s} = -\kappa_j I_j + a_j \kappa_j I_b$	$\sum_{j=0}^n I_j$	$\pi I_b - 2\pi I_b \sum_{j=0}^n a_j E_3(NYC_j x)$
	$\frac{\partial I_0}{\partial s} = 0$		
Inverse SLW	$\frac{\partial I(F_j)}{\partial s} = -\kappa_j I(F_j) + \kappa_j I_b$	$\sum_{j=0}^n I(F_j)\Delta F_j$	$\pi I_b - 2\pi I_b \sum_{j=0}^n \Delta F_j E_3\left[NYC(F_j)x\right]$
	$\frac{\partial I(F_0)}{\partial s} = 0$		
SLW-1	$\frac{\partial I_1}{\partial s} = -\kappa_1 I_1 + a_1 \kappa_1 I_b$	$I_0 + I_1$	$\pi I_b a_1 - 2\pi I_b a_1 E_3\left[NYC_1 x\right]$
	$\frac{\partial I_0}{\partial s} = 0$		
Inverse SLW-1	$\frac{\partial I(F_1)}{\partial s} = -\kappa_1 I(F_1) + \kappa_1 I_b$	$I(\tilde{F}_0)\Delta F_0 + I(F_1)\Delta F_1$	$\pi I_b - 2\pi I_b \sum_{j=0}^1 \Delta F_j E_3\left[NYC(F_j)x\right]$
	$\frac{\partial I(\tilde{F}_0)}{\partial s} = 0$		
Exact SLW	$\frac{\partial I(C)}{\partial s} = -NYCI(C) + NYC\frac{\partial F(C)}{\partial C} I_b$	$\int_0^\infty I(C) dC$	$\pi I_b - 2\pi I_b \int_0^\infty E_3(NYCx)\frac{\partial F(C)}{\partial C} dC$
Inverse Exact SLW	$\frac{\partial I(F)}{\partial s} = -NYC(F)I(F) + NYC(F)I_b$	$\int_0^1 I(F) dF$	$\pi I_b - 2\pi I_b \int_0^\infty E_3\left[NYC(F)x\right] dF$

4. The Generalized SLW Spectral Model in Non-Isothermal Media

4.1 The SLW Reference Approach

The argument of correlated spectra (sometimes called the "ideal spectrum" assumption) is an assumption of relationship between the absorption cross-section at some chosen reference state and at the local state. The reference approach consists in the following. One chooses the reference supplemental cross-sections $\tilde{C}_0^{ref}, \tilde{C}_1^{ref}, ..., \tilde{C}_n^{ref}$. Their intersection with the gas absorption cross-section at the reference state $C_\eta\left(T_g = T_{ref}, Y = Y_{ref}\right)$ generate the spectral intervals $\Delta\eta_{i,j}$. Then there exist the local supplemental cross-sections $\tilde{C}_0^{loc}, \tilde{C}_1^{loc}, ..., \tilde{C}_n^{loc}$ such that their intersection with the absorption cross-section at local state $C_\eta\left(T_g = T_{loc}, Y = Y_{loc}\right)$ generate the same spectral intervals, as shown in Fig. 9 and as described mathematically as:

$$\Delta\eta_{i,j} = \left\{\eta:\ \tilde{C}_{j-1}^{ref} \le C_\eta\left(T_g = T_{ref}, Y = Y_{ref}\right) \le \tilde{C}_j^{ref}\right\} = \left\{\eta:\ \tilde{C}_{j-1}^{loc} \le C_\eta\left(T_g = T_{loc}, Y = Y_{loc}\right) \le \tilde{C}_j^{loc}\right\}$$

It follows from the equality of the intervals of integration that for the ALBDF at the reference source temperature

$$F\left(\tilde{C}_j^{ref}, T_g = T_{ref}, T_b = T_{ref}\right) = F\left(\tilde{C}_j^{loc}, T_g = T_{loc}, T_b = T_{ref}\right), \qquad j = 0,1,...,n$$

$$F\left(C_j^{ref}, T_g = T_{ref}, T_b = T_{ref}\right) = F\left(C_j^{loc}, T_g = T_{loc}, T_b = T_{ref}\right), \qquad j = 1,2,...,n$$

or the equivalent relationships should hold for the ALBDF at the local source temperature

$$F\left(\tilde{C}_j^{ref}, T_g = T_{ref}, T_b = T_{loc}\right) = F\left(\tilde{C}_j^{loc}, T_g = T_{loc}, T_b = T_{loc}\right), \qquad j = 0,1,...,n$$

$$F\left(C_j^{ref}, T_g = T_{ref}, T_b = T_{loc}\right) = F\left(C_j^{loc}, T_g = T_{loc}, T_b = T_{loc}\right), \qquad j = 1,2,...,n$$

Then the set of local supplemental cross-sections $\tilde{C}_0^{loc}, \tilde{C}_1^{loc}, ..., \tilde{C}_n^{loc}$ and local absorption cross-sections $C_0^{loc}, C_1^{loc}, ..., C_n^{loc}$ can be found by solution of these implicit equations. In the case of the Generalized SLW spectral model they can be found explicitly (rather than by inversion of the ALBDF) with the help of the inverse ALBDF *C*(*F*) at the reference source temperature:

$$\tilde{C}_j^{loc} = C\left[F\left(\tilde{C}_j^{ref}, T_g = T_{ref}, T_b = T_{ref}\right), T_g = T_{loc}, T_b = T_{ref}\right], \qquad j = 0,1,...,n$$

$$C_j^{loc} = C\left[F\left(C_j^{ref}, T_g = T_{ref}, T_b = T_{ref}\right), T_g = T_{loc}, T_b = T_{ref}\right], \qquad j = 1,2,...,n$$

or by the alternative equations at the local source temperature:

$$\tilde{C}_j^{loc} = C\left[F\left(\tilde{C}_j^{ref}, T_g = T_{ref}, T_b = T_{loc}\right), T_g = T_{loc}, T_b = T_{loc}\right], \qquad j = 0,1,2,...,n$$

$$C_j^{loc} = C\left[F\left(C_j^{ref}, T_g = T_{ref}, T_b = T_{loc}\right), T_g = T_{loc}, T_b = T_{loc}\right], \qquad j = 1,2,...,n$$

The alternative approaches for construction of the local cross-sections $\tilde{C}_j^{loc}$ and C_j^{loc} are visualized in figure 9. One may define the gray gas reference correlation function, which relates the local gray gas absorption cross-sections with those at the reference state:

$$B_j^{loc} = \frac{C_j^{loc}}{C_j^{ref}}$$

Then the local absorption cross-sections can be found from

$$C_j^{loc} = B_j^{loc} C_j^{ref}$$

The gray gas absorption coefficients in the RTE are local variables:

$$\kappa_\eta(s) \equiv \kappa_\eta^{loc} = N^{loc} Y^{loc} C_j^{loc} = N^{loc} Y^{loc} B_j^{loc} C_j^{ref}; \qquad \kappa_0^{loc} = 0$$

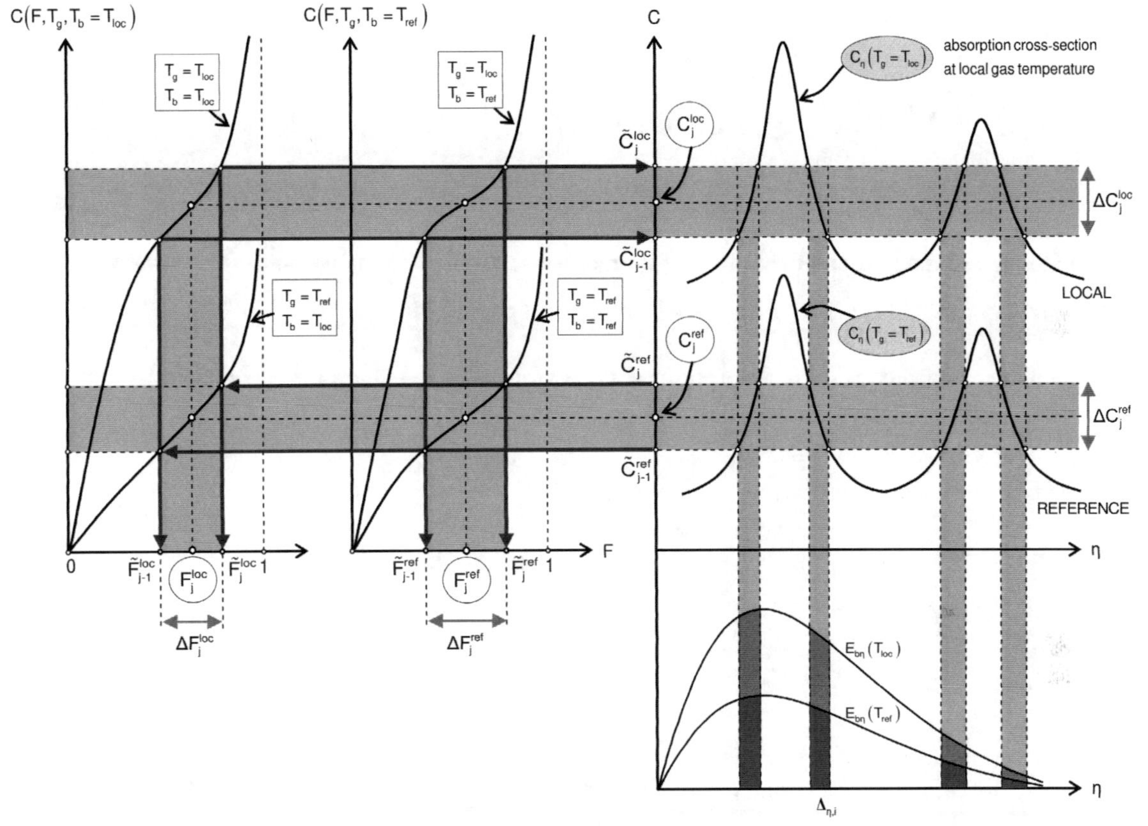

Figure 9. Correlated SLW model in non-isothermal medium – the alternative reference approaches

4.2 Outline of construction of the compact SLW model at the Reference State

The compact SLW spectral model based on Gauss-Legendre quadrature (as referenced in Section 3.2) at the reference state is constructed via the following steps.

1) Choose $T_{ref}, Y_{ref}, p_{ref}$ to define the Reference State

2) Choose n = the number of gray gases in the model

3) Define $\tilde{C}_0^{ref} = C_{min}, \ \tilde{C}_n^{ref} = C_{max}$

4) Calculate $F_{min}^{ref} = \tilde{F}_0^{ref} = F\left(C_{min}, T_g = T_{ref}, T_b = T_{ref}\right)$

$$F_{max}^{ref} = \tilde{F}_n^{ref} = F\left(C_{max}, T_g = T_{ref}, T_b = T_{ref}\right)$$

5) Calculate positive abscissas $x_j > 0$ and weights w_j, $j = 1,2,...,n$ of the Gaussian-Legendre quadratures for integration over interval $\left[-1,1\right]$.

6) Discretize the interval $\left[F_{\min}^{ref}, F_{\max}^{ref}\right]$ by transformation of the Gaussian quadratures:

$$\tilde{F}_0^{ref} = F_{min}^{ref}$$

$$\tilde{F}_j^{ref} = F_{min}^{ref} + \left(\sum_{k=1}^{j} w_k\right) \cdot \left(F_{max}^{ref} - F_{min}^{ref}\right) \qquad j = 1,2,...,n$$

$$F_j^{ref} = F_{min}^{ref} + x_j \cdot \left(F_{max}^{ref} - F_{min}^{ref}\right) \qquad j = 1,2,...,n$$

7) Calculate the reference supplemental absorption cross-sections with the help of inverse ALBDF:

$$\tilde{C}_j^{ref} = C\left(\tilde{F}_j^{ref}, T_g = T_{ref}, T_b = T_{ref}\right), \qquad j = 1,2,...,n$$

8) Calculate the reference gray gas absorption cross-sections with the help of inverse ALBDF:

$$C_j^{ref} = C\left(F_j^{ref}, T_g = T_{ref}, T_b = T_{ref}\right), \qquad j = 1,2,...,n$$

9) The finite increments of discretization become:

$$\Delta C_j^{ref} = \tilde{C}_j^{ref} - \tilde{C}_{j-1}^{ref}, \qquad j = 1,2,...,n$$

$$\Delta C_0^{ref} = \tilde{C}_0^{ref}$$

$$\Delta F_j^{ref} = \tilde{F}_j^{ref} - \tilde{F}_{j-1}^{ref}, \qquad j = 1,2,...,n$$

$$\Delta F_0^{ref} = \tilde{F}_0^{ref}$$

To determine the reference gray gas spectral intervals of integration, the entire set of wavenumbers $\left(0,\infty\right)$ is portioned into subintervals $\Delta\eta_{i,j}$ according to discretization of the continuous absorption cross-section $C_\eta\left(T_g = T_{ref}\right)$ into reference supplemental cross-sections $\tilde{C}_0^{ref}, \tilde{C}_1^{ref}, ..., \tilde{C}_n^{ref}$

$$\Delta\eta_{i,0} = \left\{\eta: \ C_\eta\left(T_g = T_{ref}\right) \le \tilde{C}_0^{ref}\right\}, \qquad \text{clear gas}$$

$$\Delta\eta_{i,j} = \left\{\eta: \ \tilde{C}_{j-1}^{ref} \le C_\eta\left(T_g = T_{ref}\right) \le \tilde{C}_j^{ref}\right\}, \qquad j = 1,2,...,n \ \text{gray gases}$$

The domain of spectral integration corresponding to gray gas j becomes

$$\Delta\eta_j = \bigcup_i \Delta\eta_{i,j}, \qquad j = 0,1,2,...,n$$

The reference spectral model is set, following which the SLW Reference Approach can be applied.

4.3 Outline of the Generalized Correlated SLW Model in Non-Isothermal Medium

1) Define the reference state by choosing T_{ref}, Y_{ref}, p_{ref}.
2) Construct the generalized spectral model at the Reference State:

$$\tilde{C}_0^{ref}, \tilde{C}_1^{ref}, ..., \tilde{C}_n^{ref}$$

$$C_1^{ref}, ..., C_n^{ref}$$

$$\tilde{F}_j^{ref} = F\left(\tilde{C}_j^{ref}, T_g = T_{ref}, T_b = T_{ref}\right), \qquad j = 0,1,2,...,n$$

$$F_j^{ref} = F\left(C_j^{ref}, T_g = T_{ref}, T_b = T_{ref}\right), \qquad j = 1,2,...,n$$

$$\tilde{F}_j^{loc} = F\left(\tilde{C}_j^{ref}, T_g = T_{ref}, T_b = T_{loc}\right), \qquad j = 0,1,2,...,n$$

$$F_j^{loc} = F\left(C_j^{ref}, T_g = T_{ref}, T_b = T_{loc}\right), \qquad j = 1,2,...,n$$

The increments of discretization are defined by:

$$\Delta C_j^{ref} = \tilde{C}_j^{ref} - \tilde{C}_{j-1}^{ref}, \qquad j = 1,2,...,n$$

$$\Delta C_0^{ref} = \tilde{C}_0^{ref}$$

$$\Delta F_j^{ref} = \tilde{F}_j^{ref} - \tilde{F}_{j-1}^{ref}, \qquad j = 1,2,...,n$$

$$\Delta F_0^{ref} = \tilde{F}_0^{ref}$$

$$\Delta F_j^{loc} = \tilde{F}_j^{loc} - \tilde{F}_{j-1}^{loc}, \qquad j = 1,2,...,n$$

$$\Delta F_0^{loc} = \tilde{F}_0^{loc}$$

3) Calculate the local absorption cross-sections and the local absorption coefficients with the help of the inverse ALBDF at the reference source temperature:

$$\tilde{C}_j^{loc} = C\left[F\left(\tilde{C}_j^{ref}, T_g = T_{ref}, T_b = T_{ref}\right), T_g = T_{loc}, T_b = T_{ref}\right], \qquad j = 0,1,...,n$$

$$C_j^{loc} = C\left[F\left(C_j^{ref}, T_g = T_{ref}, T_b = T_{ref}\right), T_g = T_{loc}, T_b = T_{ref}\right], \qquad j = 1,2,...,n$$

or by alternative equations with the help of the inverse ALBDF at the local source temperature:

$$\tilde{C}_j^{loc} = C\left[F\left(\tilde{C}_j^{ref}, T_g = T_{ref}, T_b = T_{loc}\right), T_g = T_{loc}, T_b = T_{loc}\right], \qquad j = 0,1,2,...,n$$

$$C_j^{loc} = C\left[F\left(C_j^{ref}, T_g = T_{ref}, T_b = T_{loc}\right), T_g = T_{loc}, T_b = T_{loc}\right], \qquad j = 1,2,...,n$$

Then the local gray gas absorption coefficients are

$$\kappa_\eta(s) \equiv \kappa_\eta^{loc} = \kappa_j^{loc} = N^{loc} Y^{loc} C_j^{loc}, \qquad j = 1,2,...,n$$

$$\kappa_0^{loc} = 0$$

4) Calculate the local gray gas weights by one of the following alternative approaches:

$$\Delta F_j^{loc} \equiv a_j^{loc} = F\left(\tilde{C}_j^{ref}, T_g = T_{ref}, T_b = T_{loc}\right) - F\left(\tilde{C}_{j-1}^{ref}, T_g = T_{ref}, T_b = T_{loc}\right)$$

$$\Delta F_j^{loc} \equiv a_j^{loc} = F\left(\tilde{C}_j^{loc}, T_g = T_{loc}, T_b = T_{loc}\right) - F\left(\tilde{C}_{j-1}^{loc}, T_g = T_{loc}, T_b = T_{loc}\right)$$

$$\Delta F_0^{loc} \equiv a_0^{loc} = F\left(\tilde{C}_0^{loc}, T_g = T_{loc}, T_b = T_{loc}\right)$$

$$\Delta F_0^{loc} \equiv a_0^{loc} = F\left(\tilde{C}_0^{loc}, T_g = T_{loc}, T_b = T_{loc}\right)$$

5) The gray gas RTE:

$$\frac{\partial}{\partial s} I_j(s, \mathbf{\Omega}) = -\kappa_j^{loc} I_j(s, \mathbf{\Omega}) + a_j^{loc} \kappa_j^{loc} I_b(T_{loc})$$

The alternative approches in construction of the Generalized SLW model are visualized in Fig. 9.

4.4 Additional new versions of the Reference Approach of the Generalized SLW Model

The presented alternative SLW reference approaches correspond to original SLW reference approach of Denison and Webb [5] and to the reference approach applied in the ADF method of Riviére et al. [35]. The Generalized SLW Model allows exploration of more possibilities for the reference approach: depending on starting partition into gray gases rather with the help of the supplemental cross-sections $\tilde{C}_j^{ref}$ or with the help of $\tilde{F}_j^{ref}$; depending on the choice of the blackbody source temperature rather $T_b = T_{ref}$ or $T_b = T_{loc}$ to establish gas absorption spectra correlation at different thermodynamical states; depending on the choice of calculation of the local gray gas weights a_j^{loc} with the help rather of the reference supplemental cross-sections $\tilde{C}_j^{ref}$ or with the help of the local supplemental cross-sections $\tilde{C}_j^{loc}$, totally, 8 different versions of the SLW reference approach are possible. All these possible versions are the particular cases of the visualization of the Generalized SLW model reference approach shown in Fig. 9. The full investigation and comparison of all possibilities might help to overcome the main challenge f the global methods dealing with prediction of radiative transfer in non-isothermal gaseous medium.

4.5 The finite increment form of the gray gas RTEs

One may define the following forms of the gray gas intensity

$$I_j(s, \mathbf{\Omega}) = I\left(\tilde{C}_j^{ref}, s, \mathbf{\Omega}\right) \cdot \Delta C_j^{ref} \qquad \text{The SLW gray gas increment intensity}$$

$$I_j(s, \mathbf{\Omega}) = I\left(\tilde{F}_j^{ref}, s, \mathbf{\Omega}\right) \cdot \Delta F_j^{ref} \qquad \text{The Inverse SLW gray gas increment intensity}$$

The gray gas RTEs may then be written as (with two alternative forms of the gray gas RTEs which are equivalent provided that the gas spectra are correlated):

$$\frac{\partial}{\partial s} I\left(\tilde{C}_j^{ref}, s, \mathbf{\Omega}\right) = -\kappa_j^{loc} I\left(\tilde{C}_j^{ref}, s, \mathbf{\Omega}\right) + \frac{\Delta F_j^{loc}}{\Delta C_j^{ref}} \kappa_j^{loc} I_b(T_{loc})$$

$$\frac{\partial}{\partial s} I\left(\tilde{F}_j^{ref}, s, \boldsymbol{\Omega}\right) = -\kappa_j^{loc} I\left(\tilde{F}_j^{ref}, s, \boldsymbol{\Omega}\right) + \frac{\Delta F_j^{loc}}{\Delta F_j^{ref}} \kappa_j^{loc} I_b\left(T_{loc}\right)$$

where the gray gas weights are calculated in one of two ways:

$$\begin{aligned}\Delta F_j^{loc} \equiv a_j^{loc} &= F\left(\tilde{C}_j^{ref}, T_g = T_{ref}, T_b = T_{loc}\right) - F\left(\tilde{C}_{j-1}^{ref}, T_g = T_{ref}, T_b = T_{loc}\right)\\ &= F\left(\tilde{C}_j^{loc}, T_g = T_{loc}, T_b = T_{loc}\right) - F\left(\tilde{C}_{j-1}^{loc}, T_g = T_{loc}, T_b = T_{loc}\right)\\ \Delta F_j^{ref} \equiv a_j^{ref} &= F\left(\tilde{C}_j^{ref}, T_g = T_{ref}, T_b = T_{ref}\right) - F\left(\tilde{C}_{j-1}^{ref}, T_g = T_{ref}, T_b = T_{ref}\right)\\ &= F\left(\tilde{C}_j^{loc}, T_g = T_{loc}, T_b = T_{ref}\right) - F\left(\tilde{C}_{j-1}^{loc}, T_g = T_{loc}, T_b = T_{ref}\right)\end{aligned}$$

The clear gas RTEs are described as

$$\frac{\partial}{\partial s} I_0\left(s, \boldsymbol{\Omega}\right) = 0$$

$$\frac{\partial}{\partial s} I\left(\tilde{C}_0^{ref}, s, \boldsymbol{\Omega}\right) = 0$$

$$\frac{\partial}{\partial s} I\left(\tilde{F}_0^{ref}, s, \boldsymbol{\Omega}\right) = 0$$

One solves the gray gas and clear gas RTEs subject to the boundary conditions, following which the total emissivity is calculated by summation

$$I\left(s, \boldsymbol{\Omega}\right) = \sum_{j=0}^{n} I_j\left(s, \boldsymbol{\Omega}\right)$$

$$I\left(s, \boldsymbol{\Omega}\right) = \sum_{j=0}^{n} I\left(\tilde{C}_j^{ref}, s, \boldsymbol{\Omega}\right) \cdot \Delta C_j^{ref}$$

$$I\left(s, \boldsymbol{\Omega}\right) = \sum_{j=0}^{n} I\left(\tilde{F}_j^{ref}, s, \boldsymbol{\Omega}\right) \cdot \Delta F_j^{ref}$$

4.6 Exact Limit of the Generalized SLW Model

In the limit when the number of gray gases approaches infinity, the finite increment RTEs become equations in terms of intensity as a continuous functions of the variables C and F

$$\frac{\partial}{\partial s} I\left(C, s, \boldsymbol{\Omega}\right) = -N^{loc} Y^{loc} B^{loc}\left(C\right) C I\left(C, s, \boldsymbol{\Omega}\right) + \frac{\partial F\left(C, T_g = T_{ref}, T_b = T_{loc}\right)}{\partial C} N^{loc} Y^{loc} B^{loc}\left(C\right) C I_b\left(T_{loc}\right)$$

$$\begin{aligned}\frac{\partial}{\partial s} I\left(F, s, \boldsymbol{\Omega}\right) &= -N^{loc} Y^{loc} C\left[F, T_g = T_{ref}, T_b = T_{loc}\right] I\left(F, s, \boldsymbol{\Omega}\right) + \\ &+ \frac{\partial F\left[C\left(F, T_g = T_{ref}, T_b = T_{ref}\right), T_g = T_{ref}, T_b = T_{loc}\right]}{\partial F} N^{loc} Y^{loc} C\left[F, T_g = T_{ref}, T_b = T_{loc}\right] I_b\left(T_{loc}\right)\end{aligned}$$

The total intensity is obtained by integration

$$I\left(s, \boldsymbol{\Omega}\right) = \int_0^{\infty} I\left(C, s, \boldsymbol{\Omega}\right) dC$$

$$I(s,\boldsymbol{\Omega}) \qquad = \int_0^1 I(F,s,\boldsymbol{\Omega})dF$$

4.7 Exact SLW model – Analytical Solutions based on tabulated LBL ALBDF

Calculations in the Exact SLW model can be effectively performed with the help of piecewise linear ALBDF defined by an ALBDF look-up table generated from line-by-line data. The tabulated LBL ALBDF is expressed as

$$F_j = F(C_j, T_g, T_b, Y, p) \qquad \text{interpolated to state } (T_g, T_b, Y, p), \qquad j = 1,2,...,N$$

Assuming that the ALBDF function is a piecewise linear function between the discrete entries of the look-up table

$$F(C,T_g,T_b,Y,p) = \sum_{j=1}^{N}\left[\frac{F_j - F_{j-1}}{C_j - C_{j-1}}(C - C_{j-1}) + F_{j-1}\right]\cdot\left[H(C - C_{j-1}) - H(C - C_j)\right]$$

the derivative of ALBDF in the intervals (C_{j-1}, C_j) becomes

$$\frac{\partial}{\partial C}F(C,T_g,T_b,Y,p) = \sum_{j=1}^{N}\frac{F_j - F_{j-1}}{C_j - C_{j-1}}\cdot\left[H(C - C_{j-1}) - H(C - C_j)\right]$$

Note that the derivative of ALBDF is mathematically a piecewise constant function.

4.7.1 Exact SLW modelling of Planck Mean Absorption Coefficient κ_P

Exact SLW model $$\kappa_P = NY\int_0^\infty C\frac{\partial F}{\partial C}dC = NY\sum_{j=1}^{N}(F_j - F_{j-1})\cdot\left(\frac{C_j + C_{j-1}}{2}\right)$$

SLW-*n* model $$\kappa_P = NY\sum_{j=1}^{n}(F_j - F_{j-1})\cdot\sqrt{C_j C_{j-1}}$$

4.7.2 Exact SLW modelling of the Total Emissivity ε

The Exact SLW model form of the emissivity may be stated as

$$\varepsilon = (F_{\max} - F_{\min}) - \frac{1}{NYL}\sum_{j=1}^{N}\left(\frac{F_j - F_{j-1}}{C_j - C_{j-1}}\right)\cdot\left(e^{-NYLC_{j-1}} - e^{-NYLC_j}\right)$$

which is compared to the SLW-*n* model of total emissivity as

$$\varepsilon = \sum_{j=1}^{n}(F_j - F_{j-1})\cdot\left(1 - e^{-NYLC_j}\right)$$

4.7.3 Exact SLW modelling of the net radiative flux $q(x)$

The analytical solution for the total directional radiative fluxes in a non-scattering plane layer bounded by cold, black walls may be stated as:

$$q^+(x) = \pi I_b - 2\pi I_b\int_0^\infty E_3(NYCx)\frac{\partial F}{\partial C}dC$$

$$q^-(x) = \pi I_b - 2\pi I_b \int_0^\infty E_3\left[NYC(L-x)\right]\frac{\partial F}{\partial C}dC$$

Calculation with the piecewise linearly defined ALBDF is as follows:

$$q^+(x) = \pi I_b - \frac{2\pi I_b}{NYx}\sum_{j=1}^{N}\left(\frac{F_j - F_{j-1}}{C_j - C_{j-1}}\right)\cdot\left[-E_4\left(NYC_j x\right) + E_4\left(NYC_{j-1}x\right)\right]$$

$$q^-(x) = \pi I_b - \frac{2\pi I_b}{NY(L-x)}\sum_{j=1}^{N}\left(\frac{F_j - F_{j-1}}{C_j - C_{j-1}}\right)\cdot\left\{-E_4\left[NYC_j(L-x)\right] + E_4\left[NYC_{j-1}(L-x)\right]\right\}$$

Note that the 4th order integral-exponential function appears in these equations. The exact solution may be compared to the SLW-n model result:

$$q^+(x) = \pi I_b - 2\pi I_b \sum_{j=1}^{n}\left(F_j - F_{j-1}\right)\cdot E_3\left(NYC_j x\right)$$

$$q^-(x) = \pi I_b - 2\pi I_b \sum_{j=1}^{n}\left(F_j - F_{j-1}\right)\cdot E_3\left[NYC_j(L-x)\right]$$

The total net radiative flux becomes

$$\begin{aligned} q(x) &= q^+(x) - q^-(x) \\ &= 2\pi I_b \sum_{j=1}^{n}\left(F_j - F_{j-1}\right)\cdot\left\{E_3\left[NYC_j(L-x)\right] - E_3\left(NYC_j x\right)\right\} \end{aligned}$$

4.7.4 Exact SLW Model analytical solution for the total divergence of the net radiative flux

The solution for the divergence of the net radiative flux in a plane layer bounded by cold black walls may be stated as:

$$Q(x) = -2\pi I_b(T)\int_0^\infty NYCE_2(NYCx)\frac{\partial F}{\partial C}dC - 2\pi I_b(T)\int_0^\infty (NYC)E_2\left[NYC(L-x)\right]\frac{\partial F}{\partial C}dC$$

The piecewise linear LBL ALBDF solution may likewise be stated as:

$$\begin{aligned} Q(x) = &-\frac{2\pi I_b}{x^2}\sum_{j=1}^{N}\left(\frac{F_j - F_{j-1}}{\kappa_j - \kappa_{j-1}}\right)\cdot\left[-E_4\left(\kappa_j x\right) + E_4\left(\kappa_{j-1}x\right)\right] \\ &+\frac{2\pi I_b}{x}\sum_{j=1}^{N}\left(\frac{F_j - F_{j-1}}{\kappa_j - \kappa_{j-1}}\right)\cdot\left[\kappa_j \cdot E_3\left(\kappa_j x\right) - \kappa_{j-1}\cdot E_3\left(\kappa_{j-1}x\right)\right] \\ &-\frac{2\pi I_b}{(L-x)^2}\sum_{j=1}^{N}\left(\frac{F_j - F_{j-1}}{\kappa_j - \kappa_{j-1}}\right)\cdot\left[-E_4\left((L-x)\kappa_j\right) + E_4\left((L-x)\kappa_{j-1}\right)\right] \\ &+\frac{2\pi I_b}{(L-x)}\sum_{j=1}^{N}\left(\frac{F_j - F_{j-1}}{\kappa_j - \kappa_{j-1}}\right)\cdot\left[\kappa_j \cdot E_3\left((L-x)\kappa_j\right) - \kappa_{j-1}\cdot E_3\left((L-x)\kappa_{j-1}\right)\right] \end{aligned}$$

The analytical Exact SLW Model solutions with piecewise linear interpolation of the tabulated LBL ALBDF provides the benchmark solution for validation of the SLW-*n* model.

5 The SLW-1 model

5.1 Construction of the least-squares optimized SLW-1 Model

The SLW-1 model is a minimal SLW spectral model which consists of one gray gas (κ_1, a_1), and one clear gas (κ_0, a_0), for which $\kappa_0 = 0$ and $a_0 = 1 - a_1$. Therefore, only two gray gas parameters (κ_1, a_1) are needed to define the SLW-1 model. The histogram absorption spectrum of the SLW-1 model is defined by the supplemental cross-section $\tilde{C}_0$, which separates the clear gas and the gray gas, and by absorption cross-section C_1, which defines the magnitude of the gray gas (see figure 10). The SLW-1 spectral model is not just a simple reduction of the SLW-*n* model to the case of one gray gas, *n*=1. Rather, its construction has some unique characteristics. An arbitrary SLW-1 model is not intended to be extremely accurate, although certainly not as accurate as the SLW-*n* model with sufficient gray gases. Rather, optimization is required for implementation of the SLW-1 model. The efficient SLW-1 model has been introduced previously [15,16]. These previously reported methods for construction of the SLW-1 model are simple methods which are based on exact matching of two accurately calculated radiative quantities such as the total emissivity or total flux divergence with the help of the SLW-*n* model using a large number of gray gases. These approaches are straightforward, but not robust and universal since the spectral model constructed might be sensitive to the choice of the matching parameters. The least-squares optimized SLW-1 spectral model in closed form was proposed in [17]. The method is universal, but retains its simplicity because the spectral parameters are found from explicit analytical relations. It is based on exponential approximation of the integro-exponential function widely used in radiation theory. In this approach, the objective function was defined on the entire layer for $0 \le x \le L$. In recent research [46], it has been observed that minimization of the objective function on the equivalent plane layer $0 \le x \le R$, where the reduced layer thickness $R = c \cdot L$ with some coefficient $c < 1$, might yield better results for prediction of radiative transfer in the entire layer.

Construction of the "true" least-squares optimized SLW-1 model consists in the following: consider the normalized total net radiative flux in the isothermal layer at temperature *T* bounded by cold black walls. Expressions for the dimensionless radiative flux for the SLW-*n* and SLW-1 models may be stated as

SLW-*n* model
$$\frac{q_{SLW\text{-}n}(x)}{2\pi I_b(T)} = -\sum_{j=0}^{n} a_j \left\{ E_3(\kappa_j x) - E_3\left[\kappa_j (L-x)\right] \right\}$$

SLW-1 model $$\frac{q_{SLW\text{-}1}(x)}{2\pi I_b(T)} = -a_1\left\{E_3(\kappa_1 x) - E_3\left[\kappa_1(L-x)\right]\right\}$$

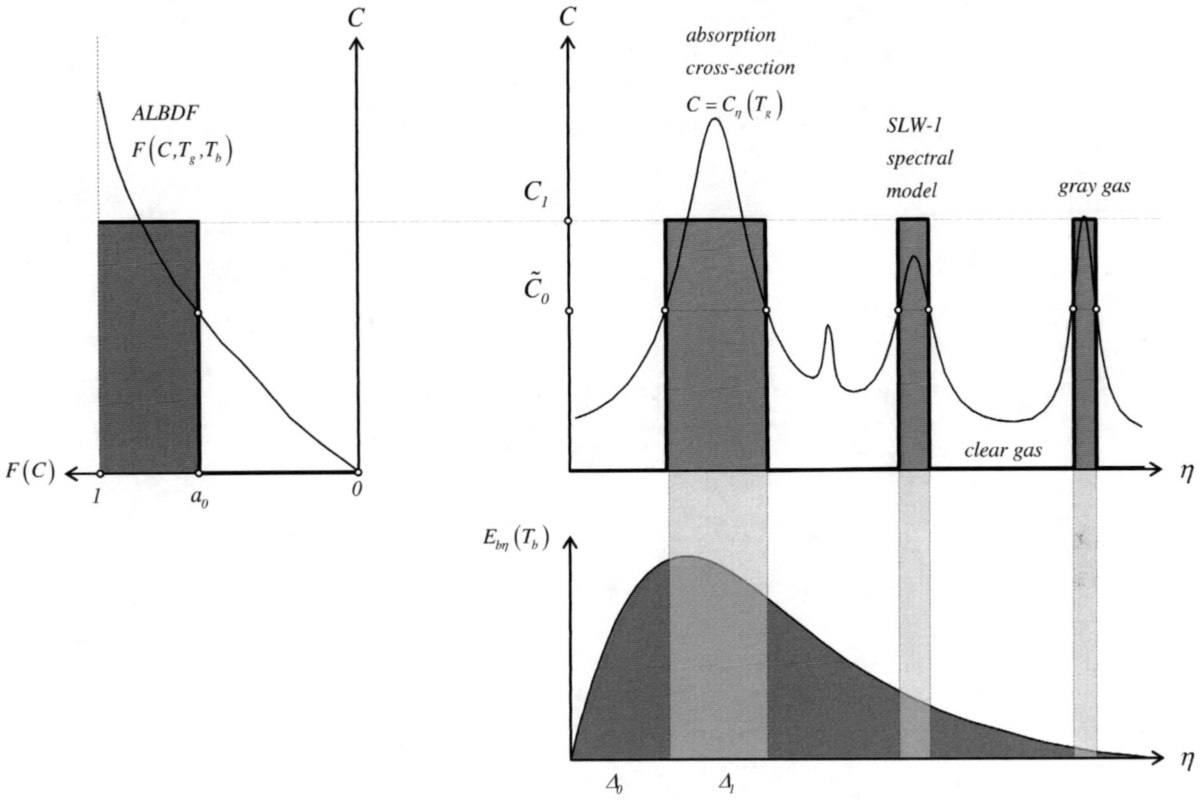

Figure 10. The SLW-1 spectral model

If one discretizes the spatial interval $0 \le x \le R$: $x_m = m \cdot R/M$, $m = 0,1,2,\ldots,M$ (*i.e.*, uniform partition) Then one may find the parameters κ_1 and a_1 of the SLW-1 model determined from minimizing the least-square error

$$P(a_1,\kappa_1) \equiv \frac{1}{2\pi I_b(T)}\sum_{m=1}^{M}\left[q_{SLW\text{-}n}(x_m) - q_{SLW\text{-}1}(x_m)\right]^2 \rightarrow min$$

The optimal solution should be at the points κ_1 and a_1 where the partial derivatives of the objective function $P(a_1,\kappa_1)$ with respect to κ_1 and a_1 vanish (*i.e.*, at the critical points):

$$\frac{\partial}{\partial a_1}P(a_1,\kappa_1) = \frac{\partial}{\partial a_1}\sum_{m=0}^{M}\left[\frac{q_{SLW\text{-}n}(x_m)}{2\pi I_b(T)} - \frac{q_{SLW\text{-}1}(x_m)}{2\pi I_b(T)}\right]^2$$

$$= \frac{\partial}{\partial a_1}\sum_{m=0}^{M}\left[\frac{q_{SLW\text{-}n}(x_m)}{2\pi I_b(T)} - a_1\left\{E_3\left[\kappa_1(L-x)\right] - E_3(\kappa_1 x)\right\}\right]^2 = 0$$

One may now differentiate and solve this equation for a_1 in terms of κ_1:

$$a_1 = \frac{2\sum_{m=1}^{M}\left[\frac{q_{SLW\text{-}n}(x_m)}{2\pi I_b(T)}\left\{E_3(\kappa_1 x_m)-E_3\left[\kappa_1(L-x_m)\right]\right\}\right]}{\sum_{m=1}^{M}\left\{E_3(\kappa_1 x_m)-E_3\left[\kappa_1(L-x_m)\right]\right\}^2}$$

Then the differentiation of the objective function with respect to κ_1 yields

$$\frac{\partial}{\partial\kappa_1}P(a_1,\kappa_1)$$

$$=2a_1\sum_{m=0}^{M}\left[\frac{q_{SLW\text{-}n}(x_m)}{2\pi I_b(T)}-a_1\left\{E_3\left[\kappa_1(L-x)\right]-E_3(\kappa_1 x)\right\}\right]\left\{(L-x_m)E_2\left[\kappa_1(L-x_m)\right]-x_m E_3(\kappa_1 x_m)\right\}=0$$

One may then replace a_1 and obtain the expression for the unknown κ_1:

$$\sum_{m=0}^{M}\left[\frac{q_{SLW\text{-}n}(x_m)}{2\pi I_b(T)}+\frac{2\sum_{m=1}^{M}\left[\frac{q_{SLW\text{-}n}(x_m)}{2\pi I_b(T)}\left\{E_3(\kappa_1 x_m)-E_3\left[\kappa_1(L-x_m)\right]\right\}\right]}{\sum_{m=1}^{M}\left\{E_3(\kappa_1 x_m)-E_3\left[\kappa_1(L-x_m)\right]\right\}^2}\left\{E_3(\kappa_1 x_m)-E_3\left[\kappa_1(L-x_m)\right]\right\}\right]\times$$

$$\times\left\{x_m E_3(\kappa_1 x_m)-(L-x_m)E_2\left[\kappa_1(L-x_m)\right]\right\} = 0$$

Finally, this equation is solved numerically for κ_1 (it has a unique solution for $\kappa_1 > 0$), and a calculation of a_1 follows. An example of the SLW-1 model is shown in figure 11, where the gray gas absorption coefficient κ_1 is found for a layer of water vapor of thickness $L = 0.75m$ and temperature of $750K$. For the case of black cold boundaries, the total net radiative flux $q(x)$, kW/m^2 is calculated with the help of the SLW-10, and with the optimized SLW-1 model (see table 1).

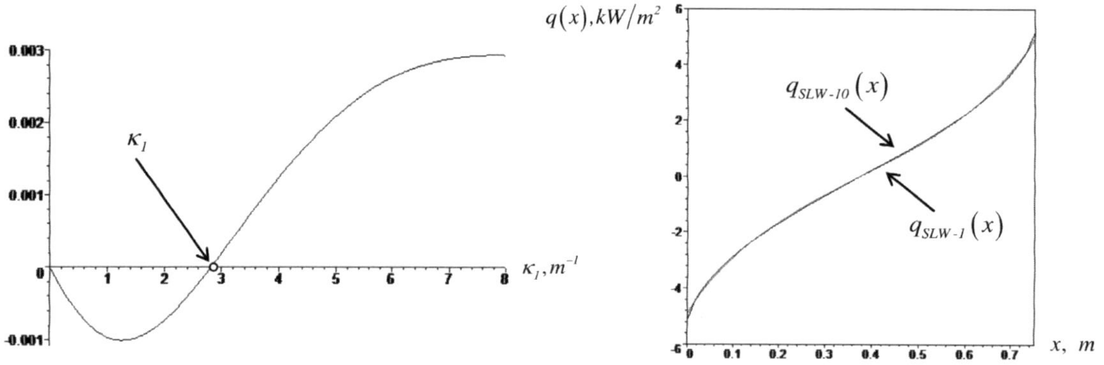

Figure 11. a) The optimized parameter κ_1; b) net radiative flux by SLW-10 and SLW-1 models

5.2 Outline of the SLW-1 Model in Non-Isothermal Medium – the Reference Approach

The development of the SLW-1 model in non-isothermal media using the Reference Approach uses the following sequence(see figure 12 for visualization):

1. Choose the reference state $T_{ref}, Y_{ref}, p_{ref}$.

   Determine the parameters of the efficient SLW-1 model at the reference state:

   κ_1^{ref}, a_1^{ref}, then calculate $\kappa_0^{ref} = 0,\ a_0^{ref} = 1 - a_1^{ref}$

2. Determine the cross-sections of the SLW-1 model at the reference state using inverse ALBDF:

$$\tilde{C}_0^{ref} = C\left(a_0^{ref}, T_g = T_{ref}, T_b = T_{ref} \right)$$

$$C_1^{ref} = \frac{\kappa_1^{ref}}{N^{ref} Y^{ref}}$$

3. Determine the absorption cross-section C_1^{loc} at the local state using inverse ALBDF:

$$C_1^{loc} = C\left[F\left(C_1^{ref}, T_g = T_{ref}, T_b = T_{ref} \right), T_g = T_{loc}, T_b = T_{ref} \right]$$

4. Calculate the SLW-1 model parameters at the local state:

$$a_0^{loc} = F\left(\tilde{C}_0^{ref}, T_g = T_{ref}, T_b = T_{loc} \right), \qquad a_1^{loc} = 1 - a_0^{loc}$$

$$a_0(T_0) = F\left(\tilde{C}_0^{ref}, T_g = T_{ref}, T_b = T_0 \right), \quad a_1(T_0) = 1 - a_0(T_0)$$

$$a_0(T_L) = F\left(\tilde{C}_0^{ref}, T_g = T_{ref}, T_b = T_L \right), \quad a_1(T_L) = 1 - a_0(T_L)$$

$$\kappa_0^{ref} = 0,\ \kappa_1^{loc} = N^{loc} Y^{loc} C_1^{loc}$$

5. RTEs for clear and gray gas:

$$\frac{\partial I_0^+(x,\mu)}{\partial x} = 0$$

$$\frac{\partial I_0^-(x,\mu)}{\partial x} = 0$$

$$\mu \frac{\partial I_1^+(x,\mu)}{\partial x} = -\kappa_1(x) I_1^+(x,\mu) + \kappa_1(x) a_1(x) I_b(T)$$

$$-\mu \frac{\partial I_1^-(x,\mu)}{\partial x} = -\kappa_1(x) I_1^-(x,\mu) + \kappa_1(x) a_1(x) I_b(T)$$

6. Boundary conditions in the case of the gray diffusely reflective boundaries are:

$$I_j^+(0,\mu) = \varepsilon_0 a_j(T_0) I_b(T_0) + 2\rho_0 \int_0^1 I_j^-(0,\mu)\mu d\mu \qquad j = 0,1$$

$$I_j^-(L,\mu) = \varepsilon_L a_j(T_L) I_b(T_L) + 2\rho_L \int_0^1 I_j^+(L,\mu)\mu d\mu \qquad j = 0,1$$

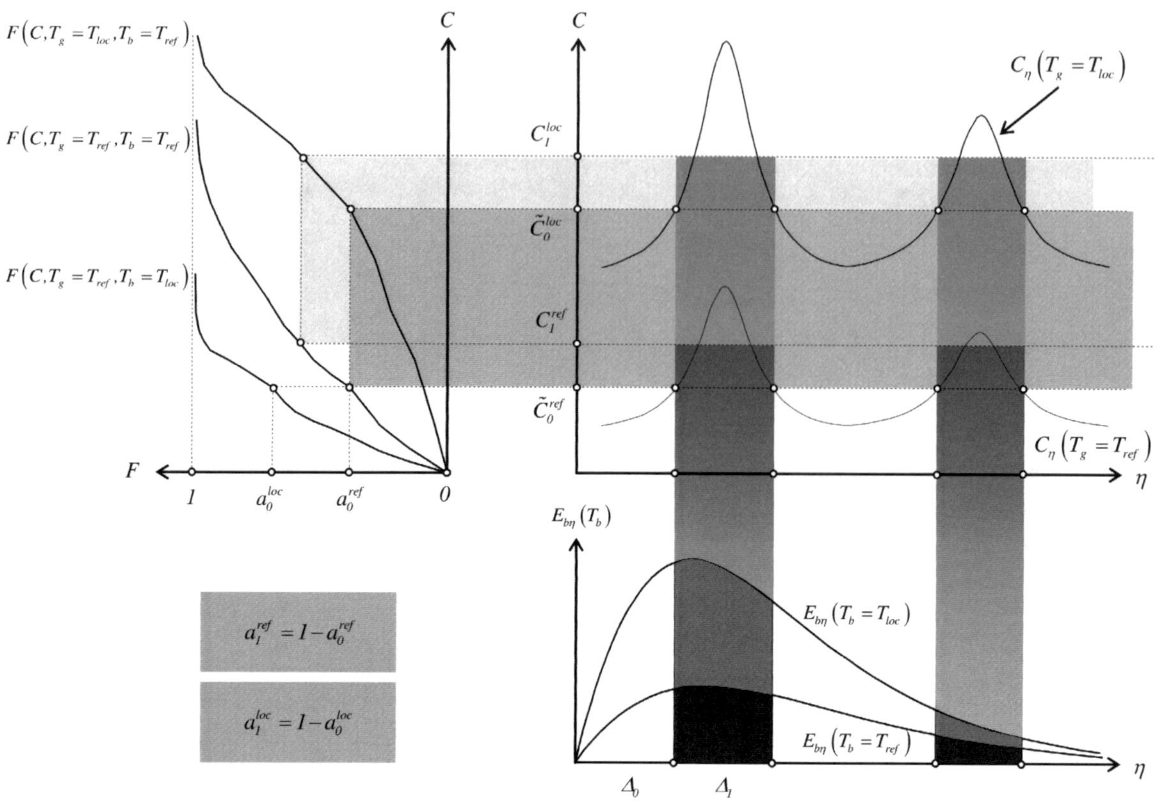

Figure 12. Reference approach for SLW-1 model

5.3 The SLW-1 Model in 2-D and 3-D geometries

Extensive investigation of different approaches to the construction of the SLW-1 model in two-dimensional rectangular and cylindrical geometries; and three-dimensional rectangular, cylindrical and spherical geometries for radiative transfer of mixtures without soot and with soot particles has been performed recently [47]. The approach taken there is identifying an equivalent plane layer in which the SLW-1 model is built with the help of least-squares optimization and then applied to higher spatial dimension. It has been observed that for most cases considered a satisfactory SLW-1 model can be constructed. However, the research is still far from a definitive conclusion and an appropriate recommendation regarding the implementation of the SLW-1 model in multidimensional situations.

6. The SLW modeling of the total emissivity, and the Planck and the Rosseland mean absorption coefficients

Total radiation quantities such as gas total emissivity or different types of averaged absorption coefficients (*e.g.*, Planck or the Rosseland mean absorption coefficients) are different characterizations of the gas absorption spectrum. They are used in some spectral models for prediction of radiative transfer, for example, in the optically thin or in the optically thick approximations. However, they can also be used for construction of spectral models which preserve these total characteristics. For example, the parameters of the SLW-*n* or SLW-1 models can be found by fitting the parameters ε, κ_P, or κ_R, evaluated by one of several methods in optically thin or optically thick media.

One must emphasize that in the Generalized SLW model, ε, κ_P, and κ_R are not total characteristics of the actual high resolution absorption spectrum but rather, the total characteristics of the gray gas absorption coefficient reordered with the help of the direct and the inverse ALBDF. They have very clear mathematical interpretations and geometrical visualization. The generalized SLW model yields a variety of the representations of the total characteristics summarized in the table 2. Note that the SLW model calculation of ε, κ_P, or κ_R involves summation only over gray gases $j = 1,2,...,n$. Therefore, these parameters are treated as characterization of the reordered gray gas absorption coefficients of the SLW model. The approach to and implications of calculating these parameters is considered in this section.

6.1 SLW Modelling of Total Emissivity

Spectral integration of the spectral emissivity with the SLW model yields the total emissivity of a gas layer [1-3]

$$\varepsilon = \frac{\int_0^\infty \varepsilon_\eta I_{b\eta}(T) d\eta}{I_b(T)} = \sum_{j=1}^{n} a_j \cdot \left(1 - e^{-\kappa_j L}\right)$$

where the gray gas weights are calculated with the help of the ALBDF:

$$a_j = F\left(\tilde{C}_j, T_g = T, T_b = T\right) - F\left(\tilde{C}_{j-1}, T_g = T, T_b = T\right)$$

Note that the clear gas does not contribute to the emissivity. All possible SLW forms of modeling of the emissivity ε are shown in table 2.

6.2 SLW Modelling of the Planck mean absorption coefficient κ_P

The SLW modeling of the Planck mean absorption coefficient yields [8,9]:

$$\kappa_P(T) = \frac{\int_0^\infty \kappa_\eta I_{b\eta}(T) d\eta}{I_b(T)} = \sum_{j=1}^{n} \kappa_j a_j = \sum_{j=1}^{n} \kappa_j \cdot \left[F\left(C_j, T_g = T, T_b = T\right) - F\left(C_{j-1}, T_g = T, T_b = T\right)\right]$$

The geometric meaning of κ_P is the area under the graph of the *F*(*C*) curve. The modified Planck mean absorption coefficient is defined by distinguishing the gas and source temperatures

$$\kappa_P\left(T_g, T_b\right) = \sum_{j=1}^{n} \kappa_j a_j = \sum_{j=1}^{n} \kappa_j \left[F\left(C_j, T_g, T_b\right) - F\left(C_{j-1}, T_g, T_b\right)\right]$$

Note that the clear gas does not contribute to κ_P. All possible SLW forms of κ_P are listed in table 2.

6.3 SLW modelling of the Rosseland Mean Absorption Coefficient κ_R

The Rosseland mean absorption coefficient appears in the diffusion approximation of radiation transfer through optically thick media, and is defined as

$$\frac{1}{\kappa_R\left(T_g, T_b\right)} = \frac{1}{\frac{\partial}{\partial T_b}\left[I_b\left(T_b\right)\right]} \cdot \int_0^\infty \frac{1}{\kappa_\eta\left(T_g\right)} \cdot \frac{\partial}{\partial T_b}\left[I_{b\eta}\left(T_b\right)\right] d\eta = \frac{\pi}{4\sigma T_b^3} \cdot \int_0^\infty \frac{1}{\kappa_\eta\left(T_g\right)} \cdot \frac{\partial}{\partial T_b}\left[I_{b\eta}\left(T_b\right)\right] d\eta$$

The term $1/\kappa_R$ may be treated as a mean free path in the absorbing medium. If the medium is optically thick over the entire wavenumber region, the Rosseland mean absorption coefficient $\kappa_R\left(T_g,T_b\right)$ exists and is calculated in a straightforward way. The Rosseland mean of the absorption coefficient of a single Lorentz line without truncation which is always non-zero exists and it can be calculated directly. Therefore, the Rosseland mean of the absorption coefficient of a gas spectrum constructed as a superposition of the Lorentz lines without truncation exists also. The problem of calculating κ_R for gases appears when the absorption spectrum due to truncation of spectral lines includes completely transparent bands, *i.e.*, where $1/\kappa_R$ is not defined. Therefore, the Rosseland mean for gases which include spectral windows does not exist. Moreover, for any gas, there are always spectral intervals where the gas is optically thin. Therefore, $1/\kappa_R$ is defined for strong bands only, where the Rosseland mean absorption coefficient refers to the mean over the optically thick bands [38,39].

The Rosseland mean absorption coefficient is defined here as the mean of the reordered absorption coefficient $\kappa(F)=NYC\left(F,T_g,T_b\right)$ defined in $F\in\left[F_{min},F_{max}\right]$ without inclusion of clear gas. Therefore, it is the Rosseland mean of the absorbing gray gases only and it characterizes the SLW spectral model rather than the actual high resolution gas absorption coefficient (which might include spectral windows). Its application has a sense only for the current SLW spectral model. The derivation of the SLW model for the term $1/\kappa_R$ is the following:

$$\frac{1}{\kappa_R\left(T_g,T_b\right)}=\frac{\int_0^\infty \frac{1}{\kappa_\eta\left(T_g\right)}\frac{\partial}{\partial T_b}E_{b\eta}\left(T_b\right)d\eta}{4\sigma T_b^3}$$

$$=\frac{\sum_{j=1}^{n}\int_{\Delta_j}\frac{1}{\kappa_\eta\left(T_g\right)}\frac{\partial}{\partial T_b}E_{b\eta}\left(T_b\right)d\eta}{4\sigma T_b^3}$$

$$=\frac{\sum_{j=1}^{n}\frac{1}{\kappa_j}\frac{\partial}{\partial T_b}\left[E_b\left(T_b\right)\cdot\left(F\left(\tilde{C}_j,T_g,T_b\right)-F\left(\tilde{C}_{j-1},T_g,T_b\right)\right)\right]}{4\sigma T_b^3}=\frac{\sum_{j=1}^{n}\frac{1}{\kappa_j}\frac{\partial}{\partial T_b}\left[E_b\left(T_b\right)\cdot a_j\right]}{4\sigma T_b^3}$$

$$=\frac{\sum_{j=1}^{n}\frac{1}{\kappa_j}\left[\frac{\partial}{\partial T_b}E_b\left(T_b\right)\cdot a_j+E_b\left(T_b\right)\cdot\frac{\partial}{\partial T_b}a_j\right]}{4\sigma T_b^3}$$

$$=\sum_{j=1}^{n}\frac{1}{\kappa_j}\left[a_j+\frac{T_b}{4}\cdot\frac{\partial}{\partial T_b}a_j\right]$$

Therefore, the SLW model representation of the Rosseland mean absorption coefficient of gray gases is given by

$$\frac{1}{\kappa_R\left(T_g,T_b\right)}=\sum_{j=1}^{n}\frac{1}{\kappa_j}\cdot\left[a_j+\frac{T_b}{4}\cdot\frac{\partial}{\partial T_b}a_j\right],\qquad a_j=F\left(\tilde{C}_j,T_g,T_b\right)-F\left(\tilde{C}_{j-1},T_g,T_b\right)$$

Under the SLW-1 modeling approach this reduces to

$$\frac{1}{\kappa_R\left(T_g,T_b\right)}=\frac{1}{\kappa_1}\cdot\left[a_1+\frac{T_b}{4}\cdot\frac{\partial}{\partial T_b}a_1\right]$$

or it can also be expressed in terms of the weight of the clear gas

$$\frac{1}{\kappa_R\left(T_g,T_b\right)} = \frac{1}{\kappa_1}\cdot\left[1-a_0-\frac{T_b}{4}\cdot\frac{\partial}{\partial T_b}a_0\right]$$

The Exact SLW Model of the Rosseland mean absorption coefficient may be derived as follows:

$$\frac{1}{\kappa_R\left(T_g,T_b\right)} = \frac{\sum_{j=1}^{n}\frac{1}{\kappa_j}\frac{\partial}{\partial T_b}\left[E_b\left(T_b\right)\cdot\left(F\left(\tilde{C}_j,T_g,T_b\right)-F\left(\tilde{C}_{j-1},T_g,T_b\right)\right)\right]}{\frac{\partial}{\partial T_b}\left[E_b\left(T_b\right)\right]}$$

$$= \frac{\sum_{j=1}^{n}\frac{1}{NY\tilde{C}_j}\frac{\partial}{\partial T_b}\left[E_b\left(T_b\right)\cdot\frac{F\left(\tilde{C}_j,T_g,T_b\right)-F\left(\tilde{C}_{j-1},T_g,T_b\right)}{\Delta C_j}\right]\Delta C_j}{\frac{\partial}{\partial T_b}\left[E_b\left(T_b\right)\right]}$$

$$= \frac{\int_0^{\infty}\frac{1}{NYC}\frac{\partial}{\partial T_b}\left[E_b\left(T_b\right)\cdot\frac{\partial}{\partial C}F\left(C,T_g,T_b\right)\right]dC}{\frac{\partial}{\partial T_b}\left[E_b\left(T_b\right)\right]}$$

$$= \frac{\frac{1}{NY}\int_0^{\infty}\frac{1}{C}\left[\frac{\partial}{\partial T_b}\left(\sigma T_b^4\right)\cdot\frac{\partial}{\partial C}F\left(C,T_g,T_b\right)+\left(\sigma T_b^4\right)\cdot\frac{\partial}{\partial T_b}\frac{\partial}{\partial C}F\left(C,T_g,T_b\right)\right]dC}{\frac{\partial}{\partial T_b}\left[\sigma T_b^4\right]}$$

As a result, the Exact SLW model of κ_R is given by

$$\frac{1}{\kappa_R\left(T_g,T_b\right)} = \frac{1}{NY}\int_0^{\infty}\frac{1}{C}\left[\frac{\partial}{\partial C}F\left(C,T_g,T_b\right)+\left(\frac{T_b}{4}\right)\cdot\frac{\partial}{\partial T_b}\frac{\partial}{\partial C}F\left(C,T_g,T_b\right)\right]dC$$

Application of the LBL look-up table of the ALBDF for calculation of the Rosseland mean absorption coefficient becomes

$$\frac{1}{\kappa_R\left(T_g,T_b\right)}$$

$$= \frac{1}{NY}\sum_{i=1}^{n}\frac{\ln\frac{C_i}{C_{i-1}}}{C_i-C_{i-1}}\left[F\left(C_i,T_g,T_b\right)-F\left(C_i,T_g,T_b\right)+\left(\frac{T_b}{4}\right)\cdot\frac{F\left(C_i,T_g,T_b\right)-F\left(C_i,T_g,T_b\right)-F\left(C_i,T_g,T_b-\Delta T\right)+F\left(C_i,T_g,T_b-\Delta T\right)}{\Delta T}\right]$$

Geometrical interpretation of the SLW and the Exact SLW modelling of mean coefficients κ_P and κ_R are shown in figure 13.

The Rosseland mean absorption coefficient of H_2O, CO_2, and CO at different gas temperature calculated with the help of correlations and look-up table for ALBDF is shown in figure 14. It should be noted that calculation of the Rosseland mean absorption coefficient (as well as of the Planck mean absorption coefficient) does not involve any geometric restrictions, and, therefore, they are the general characterization of the gas absorption spectrum.

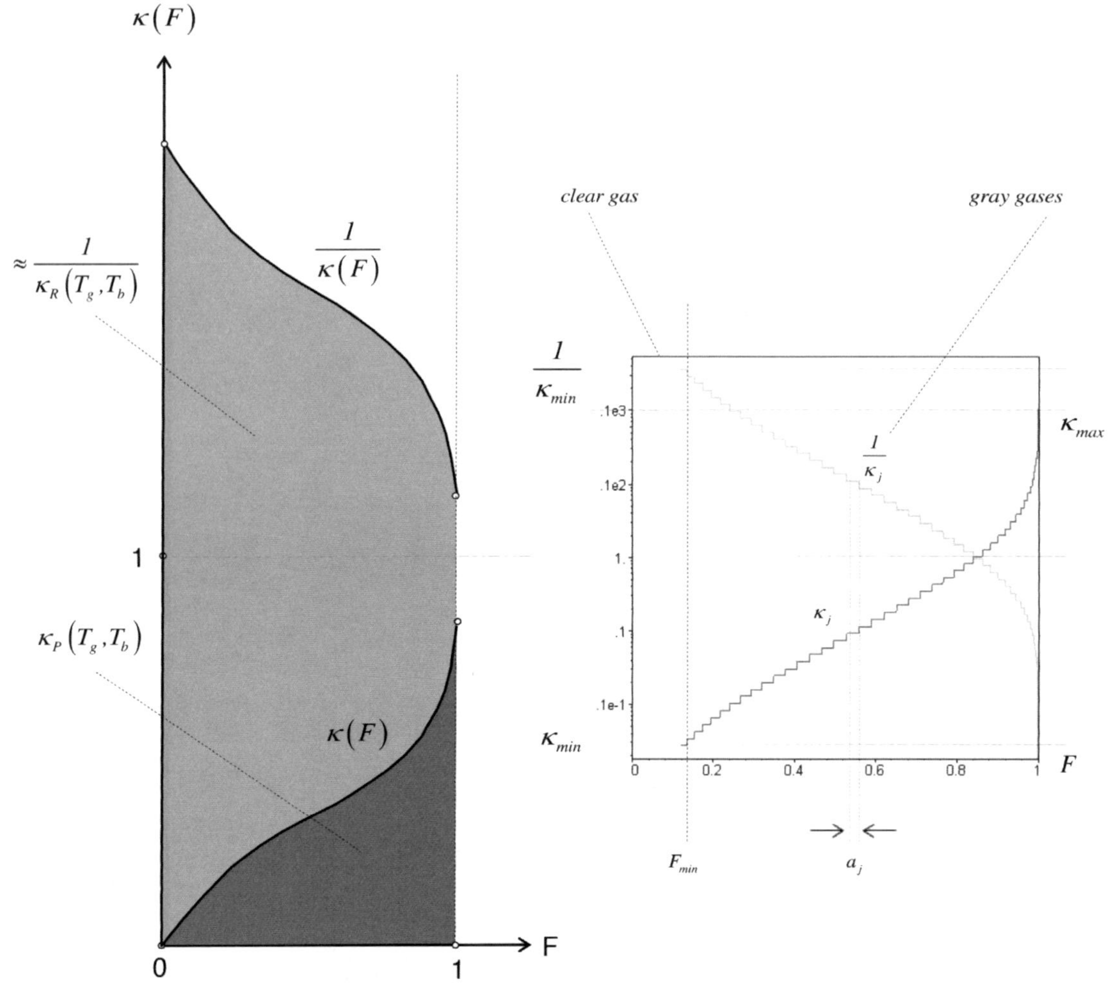

Figure 13. Reordered absorption coefficient of the Exact SLW model and geometrical interpretation of SLW modelling of κ_P and κ_R

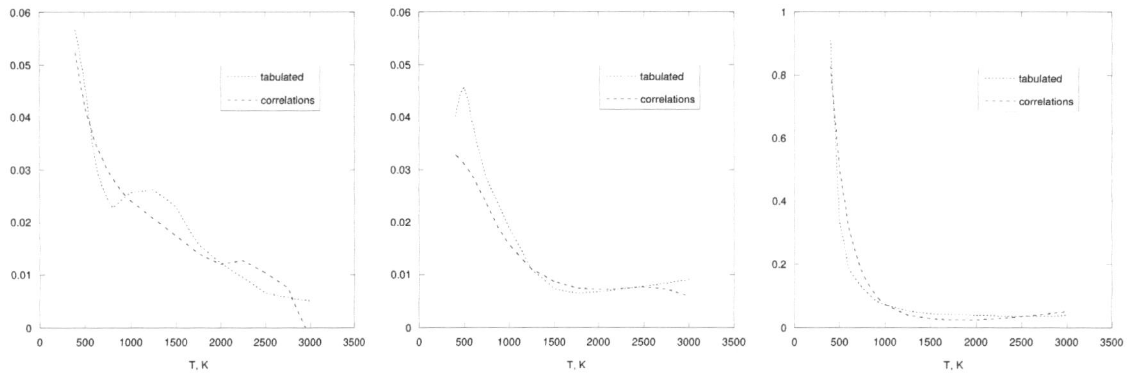

Figure 14. Dependence of the Rosseland mean absorption coefficient $\kappa_R, \left[m^{-1}\right]$ on gas temperature

a) H_2O, $Y_w = 1.0$ (self-broadening); b) CO_2, $Y_c = 1.0$ (air-broadening); c) CO, $Y_m = 1.0$ (air-broadening)

Table 2. Generalized SLW modelling of ε , κ_P and κ_R

Model	*Total Emissivity* $\varepsilon(L)$	*Planck Mean* κ_P	*Rosseland Mean* $1/\kappa_R$
LBL	$\frac{1}{I_b}\int_0^\infty\left(1-e^{-\kappa_\eta L}\right)I_{b\eta}d\eta$	$\frac{1}{I_b}\int_0^\infty \kappa_\eta I_{b\eta}d\eta$	$\frac{1}{\frac{\partial}{\partial T_b}I_b(T_b)}\cdot\int_0^\infty\frac{1}{\kappa_\eta(T_g)}\frac{\partial}{\partial T_b}\left[I_{b\eta}(T_b)\right]d\eta$
SLW	$\sum_{j=1}^{n}a_j\left(1-e^{-NYC_jL}\right)$	$\sum_{j=1}^{n}a_j\kappa_j$	$\sum_{j=1}^{n}\frac{1}{\kappa_j}\left[a_j+\frac{T_b}{4}\frac{\partial a_j}{\partial T_b}\right]$
Inverse SLW	$\sum_{j=1}^{n}\Delta F_j\left[1-e^{-NYC(F_j)L}\right]$	$NY\sum_{j=1}^{n}\Delta F_jC(F_j)$	$\sum_{j=1}^{n}\frac{1}{NYC(F_j)}\left[\Delta F_j+\frac{T_b}{4}\frac{\partial}{\partial T_b}\Delta F_j\right]$
SLW-1	$a_1\left(1-e^{-\kappa_1 L}\right)$	$a_1\kappa_1$	$\frac{1}{\kappa_1}\left[a_1-\frac{T_b}{4}\frac{\partial a_0}{\partial T_b}\right]$
Inverse SLW-1	$\Delta F_1\left[1-e^{-NYC(F_1)L}\right]$	$NY\Delta F_1C(F_1)$	$\frac{1}{\kappa_1}\left[\Delta F_1+\frac{T_b}{4}\frac{\partial\Delta F_1}{\partial T_b}\right]$
Exact SLW	$\int_0^\infty\left(1-e^{-NYCL}\right)\frac{\partial F(C)}{\partial C}dC$ $1-NYL\int_0^\infty e^{-NYCL}F(C)dC$	$NY\int_0^\infty C\frac{\partial F(C)}{\partial C}dC$ $NY\int_0^\infty\left[1-F(C)\right]dC$	$\frac{1}{NY}\int_0^\infty\frac{1}{C}\left[\frac{\partial F(C)}{\partial C}+\frac{T_b}{4}\frac{\partial^2F(C,T_g,T_b)}{\partial T_b\partial C}\right]dC$
Inverse Exact SLW	$1-\int_0^1 e^{-NYC(F)L}dF$	$NY\int_0^1 C(F)dF$	$\approx\frac{1}{NY}\int_0^1\frac{dF}{C(F)}$

Journal of Physics: Conference Series **676** (2016) 012022 doi:10.1088/1742-6596/676/1/012022

Comment regarding application of the κ_P and κ_R for construction of the semi-gray model
If the term with the derivative of weight with respect to source temperature is neglected, then the simplified SLW-1 model representation of κ_R is given by

$$\frac{1}{\kappa_R\left(T_g,T_b\right)} = \frac{1}{\kappa_1}\cdot\left[a_1+\frac{T_b}{4}\cdot\frac{\partial}{\partial T_b}a_1\right] \approx \frac{a_1}{\kappa_1}$$

Recall also the SLW-1 model representation of the Planck mean absorption coefficient is

$$\kappa_P = a_1\kappa_1$$

These two equations can be used for determining the parameters of the SLW-1 model in terms of κ_P and κ_R:

$$\kappa_1=\sqrt{\kappa_P\kappa_R} \text{ and } a_1=\sqrt{\kappa_P/\kappa_R}$$

These quantities which are the parameters of the SLW-1 model based on κ_P and κ_R are also known, $\kappa_1=\sqrt{\kappa_P\kappa_R}$ as the mean absorption coefficient, and $a_1=\sqrt{\kappa_P/\kappa_R}$ as the weighted nongrayness [45]. It is interesting to note that application of the SLW-1 model for κ_P and of the simplified SLW-1 model for κ_R yields also the popular semi-gray model RTE previously published in [40-42]:

$$\frac{\partial}{\partial s}I = -\sqrt{\kappa_P\kappa_R}\cdot I+\kappa_P\cdot I_b$$

However, it should be mentioned that the main difference between the SLW-1 model and the semi-gray model is that the SLW method also includes an additional equation for the clear gas.

7. Conclusions

A unified approach explores and reveals more possibilities of the SLW method, exploring both the direct and inverse implementations of the model. The present work outlines the detailed steps in the construction of the generalized SLW model in isothermal and non-isothermal media, and compares different forms of the modelled radiative quantities: directional total radiative flux, total emissivity, Planck mean and Rosseland mean absorption coefficients. Unification of both spectral models allows the resolution of some difficulties which appear when the methods are applied individually, such as inversion of the distribution function or solution of the implicit equations in applications to non-isothermal gas media. The present work can also serve as a handbook in different approaches of the SLW method.

8. Acknowledgements

This work was partially supported by the Poitou-Charentes Regional Council and by the French National Research Agency under the grant ANR-12-BS09-0018 SMART-LECT.

Journal of Physics: Conference Series **676** (2016) 012022 doi:10.1088/1742-6596/676/1/012022

9. References

[1] Denison MK 1994 *The Spectral-Line Weighted-Sum-of-Gray-Gases Model for Arbitrary RTE Solvers*, PhD Dissertation, Brigham Young University, Provo, Utah

[2] Denison MK, Webb BW 1993 A spectral line based weighted-sum-of-gray-gases model for arbitrary RTE solvers, *ASME J. Heat Transfer* **115** 1004-1012

[3] Denison MK, Webb BW 1993 An absorption-line blackbody distribution function for efficient calculation of total gas radiative transfer *J. Quant. Spectr Rad. Transfer* **50** 499-510

[4] Denison MK, Webb BW 1994 *k*-Distributions and Weighted-sum-of-gray-gases – a Hybrid Model, *ASME J.Heat Transfer* **2** 19-24

[5] Denison MK, Webb BW 1995 The spectral line based weighted-sum-of-gray-gases model in non-isothermal non-homogeneous media *ASME J. Heat Transfer* **117** 359-365

[6] Denison MK, Webb BW 1995 Development and Application of an Absorption-line Blackbody Distribution Function for CO_2 *Int. J. Heat Mass Transfer* **38** 10 1813-1821

[7] Denison MK, Webb BW 1995 The spectral-line weighted-sum-of-gray-gases model for H_2O/CO_2 Mixtures *ASME Journal of Heat Transfer* **117** 788-792

[8] Solovjov VP 1998 *Spectral Line-Based Weighted-Sum-of-Gray-Gases Modeling of Radiative Transfer in Multicomponent Mixtures of Hot Gases*, PhD Dissertation, Brigham Young University, Provo, Utah

[9] Solovjov VP, Webb BW 1998 Radiative Transfer Model Parameters for Carbon Monoxide at High Temperature, *J. Proc.11th Int. Heat Transfer Conf.* Kyongju, Korea, **7** 445-450

[10] Solovjov VP, Webb BW 2000 SLW Modeling of Radiative Transfer in Multicomponent Gas Mixtures, *J. Quant. Spectr. Rad. Transfer*, **65** 655-672

[11] Solovjov VP, Webb BW 2001 An Efficient Method for Modeling of Radiative Transfer in Multicomponent Gas Mixtures with Soot Particles, *ASME J. Heat Transfer*, **123** 450-457

[12] Solovjov VP, Webb BW 2005 The Cumulative Wavenumber Method for Modeling Radiative Transfer in Gas Mixtures with Soot, *J. Quant. Spectr Rad. Transfer* **93** 273-287

[13] Solovjov VP, Webb BW 2008 Multilayer Modeling of Radiative Transfer by SLW and CW Methods in Non-isothermal Gaseous Media, *J. Quant. Spectr Rad. Transfer* **109** 245-257

[14] Solovjov VP, Webb BW 2011 Global Spectral Methods in Gas Radiation: The Exact Limit of the SLW Model and its Relationship to the ADF and FSK Methods, *ASME J. Heat Transfer* **133** 7 88-798

[15] Solovjov VP, Lemonnier D, Webb BW 2011 The SLW-1 Model for Efficient Prediction of Radiative Transfer in High Temperature Gases *J. Quant. Spectr. Rad. Transfer* **112** 1205-12

[16] Solovjov VP, Lemonnier D, Webb BW 2011 SLW-1 Modeling of Radiative Heat Transfer in Nonisothermal Nonhomogeneous Gas Mixtures with Soot," *ASME J. Heat Transfer,* **133** 102701-3-9

[17] Solovjov VP, Lemonnier D, Webb BW 2013 Efficient Cumulative Wavenumber Method – the CW-1 Model of Radiative Transfer in the Gaseous Medium Bounded by Non-Gray Walls, *J. Quant. Spectr. Rad. Transfer* **128** 2-9

[18] Solovjov VP, Lemonnier D, Webb BW 2014 Extension of the Exact SLW model to non-isothermal gaseous media, *J. Quant. Spectr. Rad. Transfer*, **143** 83-91

[19] Pearson JT, Webb BW, Solovjov VP, Ma J 2014 Updated Correlation of the Absorption Line Blackbody Distribution Function for H_2O Based on the HITEMP2010 Database *J. Quant. Spectr. Rad. Transfer* **128** 10-17

[20] Pearson JT, Webb BW, Solovjov VP, Ma J 2014 Effect of total pressure on the absorption line blackbody distribution function and radiative transfer in H_2O, CO_2, and CO *J. Quant. Spectr. Rad. Transfer* **143** 100-110

[21] Pearson JT, Webb BW, Solovjov VP, Ma J 2014 Efficient representation of the absorption line blackbody distribution function for H_2O, CO_2, and CO at variable temperature, mole fraction, and total pressure *J. Quant. Spectr. Rad. Transfer* **138** 82-96

[22] Andre F, Vaillon R 2009 A database for the SLMB modelling of the full spectrum radiative properties of CO_2 *Comp. Thermal Rad. in Participating Media III, Proc. Of Eurotherm Seminar 83* 213 Lisboa, Portugal

[23] Solovjov VP, André F, Lemonnier D, Webb BW 2014 Le modele SLW généralisé *SFT Congrès*, Lyon

[24] André F, Solovjov VP, Vaillon R, Denis Lemonnier 2013 Modelling the Cumulative Distribution of Absorption Coefficients of Gases Using the Generalized k-Moment Method *J. Quant. Spectr. Rad. Transfer* **124** 49-61

[25] André F, Solovjov VP, Hou L, Vaillon R 2014 The Generalized *k*-moment method for the modeling of cumulative *k*-distributions of H_2O at high temperature *J. Quant. Spectr. Rad. Transfer* **143** 92-99

[26] Węcel G, Ostrovsky Z, Bialecki R 2009 A novel approach of evaluating absorption line Blackbody distribution function *Comp. Thermal Rad. in Participating Media III, Proc. of Eurotherm Seminar 83* 293-312 Lisboa, Portugal

[27] Liu F, Chu H, Zhou H, Smallwood GJ 2014 Evaluation of the Absorption Line Blackbody Distribution Function of CO_2 and H_2O Using the Proper Orthogonal Decomposition and Hyperbolic Correlations *J. Quant. Spectr. Rad. Transfer.*

[28] Chu H, Liu F, Zhou H 2011 Calculations of gas thermal radiation transfer in one-dimensional planar enclosure using LBL and SNB models *Int. J. Heat Mass Transfer* **54** 4736-45

[29] Modest MF 2013 *Radiative Heat Transfer*, Academic Press, New York, 3rd ed

[30] Modest M F. The treatment of Nongray Properties in Radiative Heat Transfer: from past to present *ASME J. Heat Transfer* **135** 061801-1-12 2013

[31] Modest M F, Zhang H 2000 The Full-Spectrum Correlated-*k* Distribution and Its Relationship to the Weighted-Sum-of-Gray-Gases Method *Proc. of the 2000 IMECE* **HTD-366-1** Orlando FL ASME New York 75–84

[32] Modest M F and Zhang H 2002 The full-spectrum correlated-*k* distribution for thermal radiation from molecular gas-particulate mixtures *ASME J. Heat Transfer* **124** 30-38

[33] Zhang H, Modest, MF 2002 A multi-scale full-spectrum correlated-k distribution for radiative heat transfer in inhomogeneous gas mixtures *J.Quant. Spectr. Rad. Transfer* **73** 349–360

[34] Pierrot L, Rivière P, Soufiani A and Taine J 1999 A ficticious-gas-based absorption distribution function global model for radiative transfer in hot gases *J. Quant. Spectr. Rad. Transfer* **62** 609

[35] Rivière P, Soufiani A, Perrin M Y, Riad H, Gleizes A. 1996 Air mixture radiative property modeling in the temperature range 10,000-40,000 K. *J. Quant. Spectr. Rad. Transfer* **56** 29-45

[36] Pierrot L, Rivière P, Soufiani A, Taine J, 1999 A fictitious-gas-based absorption distribution function global model for radiative transfer in hot gases *J. Quant. Spectr. Rad. Transfer* **62** 609–624

[37] Pierrot L, Soufiani A, Taine J 1999 Accuracy of narrow-band and global models for radiative transfer in H_2O, CO_2, and H_2O CO_2 mixtures at high temperature, *J. Quant. Spectr. Rad. Transfer* **62** 523–548

[38] Abu-Romia MM, Tien CL 1967 Appropriate Mean Absorption Coefficients for Infrared Radiation of gases *ASME J.Heat Transfer* 321-327

[39] Tien CL 1969 Thermal Radiation Properties of Gases *Advances in Heat Transfer* **5**

[40] Viskanta R 1964 Concerning the definitions of the mean absorption coefficient *Int. J. Heat Mass Transfer* **7** 1047-1049

[41] Sampson DH, Marton RT 1972 On a semi-gray approximation to non-gray radiative transfer *J. Quant. Spectr. Rad. Transfer* **12** 1389-1408

[42] Traugott SC 1968 On gray absorption coefficients in radiative transfer *J. Quant. Spectr. Rad. Transfer* **8** 971-999

[43] Rothman L S, Gordon I E, Barber R J, Dothe H, Gamache R R, Goldman A, Perevalov V I, Tahskun S A and Tennyson J 2010 HITEMP, the high-temperature molecular spectroscopic database *J. Quant. Spectr. Rad. Transfer* **111** 2139-50.

[44] Solovjov VP 1992 Solution of the Radiative Transfer Equation by the Moment Method using polynomials od special form *J. Quant. Spectr. Rad. Transfer* **47** 229-236

[45] Arpaci VS, Yang W-M 1991 Radiation affected Benard problem *in Fundamentals of Radiation Heat Transfer ASME HDT-Vol* **160** 117-126

[46] Rustam A 2014 Mise and oevre et evaluation d'un modele reduit de rayonnement des gaz en geometrie multidimensionnelle *Rapport de Stage Ingenier de l'ISAE-ENSMA 86961 Futuroscope Chasseneuil Cedex, France*

[47] Andre F, Solovjov VP, Eymet V, Caliot C 2014 Rayonnement des gaz : des spectres de raies aux modéles approchés *Ecole Thematique CNRS Four Solaire d'Odeillo*

Transition from the incoherent to the coherent regime for propagative-wave based thermal radiation

Y Tsurimaki[1,2], P-O Chapuis[1], J Okajima[2], A Komiya[2], S Maruyama[2], R Vaillon[1]
1 Université de Lyon, CNRS, INSA-Lyon, UCBL, CETHIL, UMR5008, F-69621 Villeurbanne, France
2 Institute of Fluid Science, Tohoku University, 2-1-1, Katahira, Aoba-ku, Sendai, 980-8577, Japan

E-mail: rodolphe.vaillon@insa-lyon.fr

Abstract. The transition from the incoherent to the coherent regime for thermal radiation between bodies trough a transparent medium is discussed. The canonical case of two parallel semi-infinite planar media is used as a basis to provide an insight into the physics and quantities ruling the distance at which coherent effects have an impact on the propagative component of the net heat flux exchanged. A practical criterion is proposed to define the distance below which radiation intensity framework should not be used, but instead fluctuational electrodynamics.

1. Introduction

Planck already stated that his theory would be valid provided all dimensions should be large compared to the wavelength under consideration [1]. This means that if the distance between bodies exchanging thermal radiation, or if the sizes of the bodies themselves are comparable to the wavelength, another theory is required. Fluctuational electrodynamics highlighted by Rytov and colleagues [2] provides the tool to model electromagnetic thermal radiation at small scales. Much attention has been paid to the strong enhancement of the net heat flux beyond the blackbody configuration. It gave rise to the advent of near-field thermal radiation as a new branch of radiative heat transfer (see two recent reviews and references therein [3,4]). However, much less analyses on the regime where the coherent effects of thermal radiation play a role have been proposed. It must be emphasized that this transition regime between the far-field and near-field regimes was observed in some simulation results since the early work of Polder and Van Hove [5], especially when the variations of the flux between the bodies is analyzed as a function of their separation distance. It was understood only recently that the net heat flux could be reduced far below the far-field value [6-9], and almost suppressed in some cases.

In this article, an in-depth analysis of the physics and parameters ruling the propagative component of the net radiation heat flux between semi-infinite plane-parallel plates derived from fluctuational electrodynamics is proposed. In particular, the distance between bodies at which thermal radiation cannot be considered incoherent, i.e. modelled using the intensity, is determined. Its dependence on electromagnetic properties and temperatures of the plates is discussed.

Journal of Physics: Conference Series **676** (2016) 012023 doi:10.1088/1742-6596/676/1/012023

2. Theory, results and discussion

In the case of two semi-infinite plane-parallel bodies (media 1 and 3 on figure 1.a) separated by a vacuum gap (medium 2), the formulation of the total radiative net heat flux exchanged by the two bodies was given by Polder and Van Hove [5]. This flux is split into two components, one standing for the evanescent waves and the other for the propagative waves. They are solely function of the distance between the plates (d), temperatures (T_1 and T_3) and complex permittivities (ε_1 and ε_3) of the plates, assuming that materials are non-magnetic. A radiative heat transfer coefficient can be introduced as the temperature derivative of the flux and split into the evanescent and propagative components as well:

$$h^{\text{total}}(d,T) = \lim_{\Delta T \to 0} \frac{q}{\Delta T} = \lim_{\Delta T \to 0} \frac{q^{\text{prop}}}{\Delta T} + \lim_{\Delta T \to 0} \frac{q^{\text{evan}}}{\Delta T} = h^{\text{prop}} + h^{\text{evan}} \tag{1}$$

where superscript 'total' means that propagative and evanescent waves components are both taken into account. By doing so, radiative heat transfer can be analysed as a function of the product of temperature and distance $T.d$; this is more general than an analysis as a function of the distance only for several temperature differences between the plates.

When reflectivities of the bodies are large enough, i.e. for metals, and for a symmetric configuration, it was observed previously [6-9] that the total net heat flux can decrease below the far-field value when distance d is becoming smaller. The flux can reach a minimum and eventually rise up far beyond the blackbody limit because of the contribution of evanescent waves (figure 1.b). Distance $d_{\text{incoh-coh}}$ at which the far-field radiation incoherent regime does not hold anymore for propagative waves is under investigation in the following.

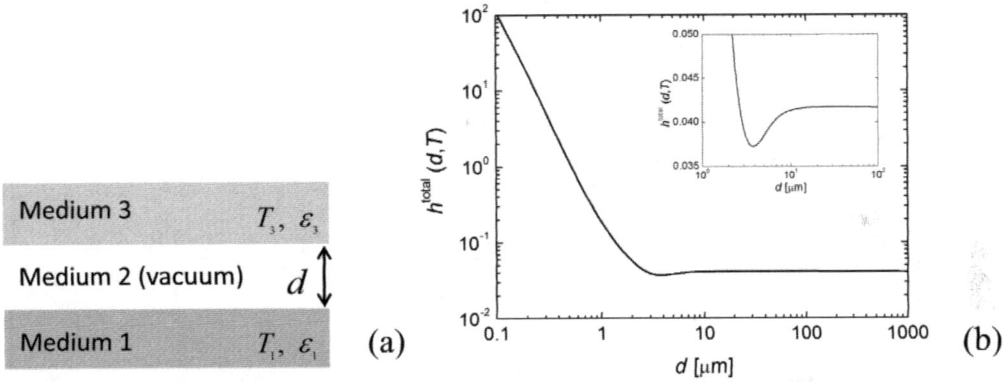

Figure 1. (a) Configuration under consideration. (b) Total radiative heat transfer coefficient as a function of distance between the plates for aluminum at 300 K.

A practical criterion is proposed to define the distance at which the transition from the far-field incoherent to the coherent regime for thermal radiation is taking place. It consists in stating that this is the distance $d_{incoh\text{-}coh}$ at which the flux has varied by 5%: $h^{\text{prop}}(d_{\text{incoh-coh}},T)/h^{\text{prop}}_{d\to\infty}(T) = 0.95$. On figure 2, this limit is plotted on a ($T.d$, T) diagram for several materials and temperatures. For comparison, the constant 2897.8 μmK from Wien's law is indicated. The wavelength where Planck's function is maximum (λ_{max}), is also known to be close to the coherence length of a blackbody emitter [10]. When the plates are not black but grey (constant relative permittivity, here equal to ε_r = 100 + i 10^{-4}), the product $T.d_{\text{incoh-coh}}$ is found to be the same at any temperature, and is smaller than $\lambda_{\text{max}}.T$. However, when real materials are considered, i.e. metals, the product $T.d_{\text{incoh-coh}}$ is much larger (around two to three times) and varies with materials and temperatures. These cases can be understood as being spectrally weighted sums, the weights being the spectral variations of permittivities and of Planck's

(Bose-Einstein) distribution function of modes. The analysis in terms of coherence length is not straightforward. The criterion of 5% variation does not necessarily coincide with any "physical" definition of what would be a coherence length of thermal radiation but it provides a limit which has a practical usefulness to state whether the far-field incoherent thermal radiation theory can be used or not for the propagating waves. This analysis will be completed by adding the cases of dissymmetric and dielectric materials. Finally, we underline that this transition length is also compared with the one associated to the transition from the propagative-wave to the evanescent-wave regime, that was not tackled here for the sake of brevity.

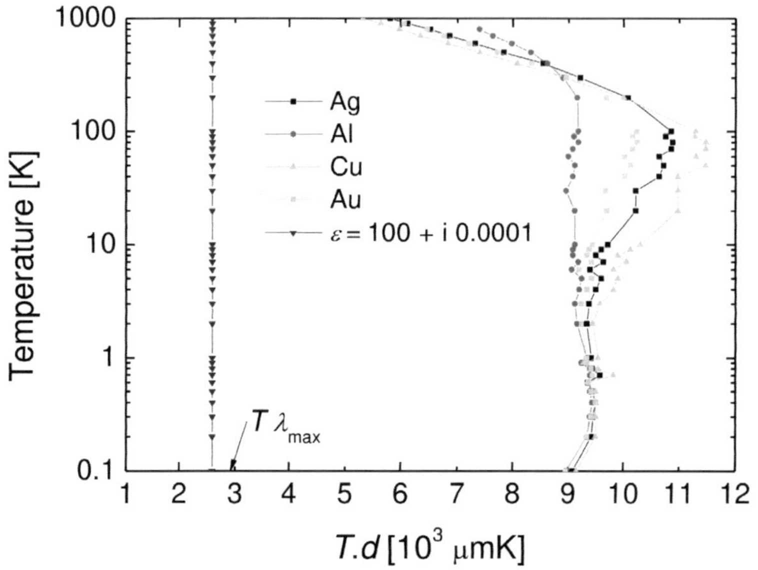

Figure 2. Plot on a ($T.d$, T) diagram of the limits $d_{\text{incoh-coh}}$ at which the transition from the far-field incoherent to coherent regime for thermal radiation takes place, for grey and metallic plates.

Acknowledgments
This research is supported by ELyT Lab (Engineering and Science Lyon – Tohoku Laboratory).

[1] Planck M 1991 *The Theory of Heat Radiation* (Courier Dover publications)
[2] Rytov SM and Kravtsov YA Tatarskii VI 1989 *Principles of Statistical Radiophysics* (*Wave Propagation Through Random Media* vol 4)
[3] Jones AC, O'Callahan BT, Yang HU and Raschke MB 2013 *Prog. Surf. Sci.* **88** 349-392
[4] Park K and Zhang Z 2013 *Front. Heat Mass Transfer* **4** 013001
[5] Polder D and Van Hove M 1971 *Phys. Rev. B* **4** 10
[6] Tsurimaki Y, Chapuis PO, Vaillon R, Okajima J, Komiya A and Maruyama S 2014 Radiative heat transfer between two semi-infinite parallel plates in the far-to-near field transition regime, *Proc. 2nd international Workshop on Micro and Nano Thermal Radiation*
[7] Mayo J and Narawanaswamy A 2014 A Minimum radiative transfer between two metallic planar surfaces *Proc. 11th AIAA/ASME Joint Thermophysics and Heat Transfer Conference*
[8] Tsurimaki Y, Chapuis PO, Vaillon R, Okajima J, Komiya A and Maruyama S 2014 Reducing thermal radiation between parallel plates in the far-to-near field transition regime *Proc. 15th International Heat Transfer Conference*
[9] Mayo J, Tsurimaki Y, Chapuis PO, Okajima J, Komiya A, Maruyama S, Narawanaswamy A and Vaillon R 2014 *Proc. Eurotherm 103 – Nanoscale and Microscale heat Transfer IV*
[10] Donges A 1998 *Eur. J. Phys.* **19** 245-249

Meshless method for solving coupled radiative and conductive heat transfer in refractive index medium

Cheng-An Wang[1], **Hamou Sadat**[2] **and Jian-Yu Tan**[1]
[1] School of Automobile Engineering, Harbin Institute of Technology at Weihai, 2 West Wenhua Road, Weihai 264209, People's Republic of China
[2] Institut PPRIME - CNRS - Université de Poitiers – ENSMA, Département Fluides, Thermique, Combustion - Campus Sud 40, avenue du recteur Pineau - 86022 POITIERS Cedex – France

E-mail: tanjianyu@hitwh.edu.cn; hamou.sadat@univ-poitiers.fr

Abstract. A diffuse approximation meshless method (DAM) is employed as a means of solving the coupled radiative and conductive heat transfer problems in semi-transparent refractive index media contained in 1D and 2D geometries. The meshless approach for radiative transfer is based on the discrete ordinates equation. Cases of combined conduction-radiation are presented, including plane parallel slab, square enclosure, and semicircular enclosure with an inner circle. The influence of the refractive index on the temperature distributions and heat fluxes is investigated. Results obtained using the proposed meshless method are compared with those reported in the literature to demonstrate the flexibility and accuracy of the method.

1. Introduction

The coupled radiative and conductive heat transfer problems in semi-transparent media at high temperature arises in many engineering applications such as nuclear engineering, solar energy collectors, cooling of electronic components, chemical reactors and so on [1,2]. Most studies on coupled radiative and conductive heat transfer problem have been limited to uniform refractive index media and employed finite element [3,4] and control volume based finite element methods [5].

When the refractive index of a semi-transparent medium varies continuously, the curved ray path is determined by the Fermat's principle. As a result, the radiative energy field inside the medium is modified. Because of the complexity of solving radiative transfer in refractive index medium, the coupled radiative and conductive heat transfer problem in the refractive index medium received relatively less attention. The curved ray tracing technique [6-8], pseudo-source adding method in combination with the curved ray tracing technique [9] or ray-tracing/nodal-analyzing method [10-12] are mainly employed for analyzing the radiative heat transfer, and traditional methods such as finite difference method or control volume are used for solving energy equation. Because of a large number of rays to be launched, the ray tracing method is complex, time consuming and difficult to settle. Mishra et al. [13] used Lattice Boltzmann method (LBM) to solve the energy equation and discrete transfer method (DTM) to compute the radiative transfer. Zhu et al. [14] employed P_N method to solve radiative transfer and spectral method to implement the spatial discretization. While DTM or P_N method is not easy to be extended to multidimensional geometries. In order to overcome these difficulties, some researchers proposed the discrete ordinates method (DOM) for the discretization of angular space of radiative transfer equation (RTE) in refractive index media. The DOM is firstly proposed by Lemonnier et al. [15] for 1D radiative heat transfer in the slab. Afterwards, another kind of DOM was derived by Liu et al. in 3D Cartesian [16] and cylindrical [17] coordinate systems

[1] To whom any correspondence should be addressed

respectively. Based on the technique of DOM, finite volume method (FVM) [16], finite element method (FEM) [18] and spectral collocation method [19] have been used for the spatial discretization.

In this article, we employ meshless method to solve radiative and conductive heat transfer in refractive index media. Different from traditional methods like FVM or FEM, the meshless method uses a set of nodes scattered in the computational spatial domain without the need for information on the relationship between them, which provides more flexibility in complex geometries [20]. To our best knowledge, only Liu [21] has used a meshless local Petrov–Galerkin (MLPG) approach for solving the coupled radiative (based on DOM) and conductive heat transfer in a one-dimensional refractive index slab. But numerical integration is necessary in this approach which increases the CPU time generally. In this paper we extend the use of DAM proposed by Nayroles [22], which has been used successfully for radiative heat transfer in refractive index media [23] and radiation coupled with conduction in a homogenous media [24]. The article is organized as following. The details of the theoretical and numerical calculation procedures are given first. To evaluate the accuracy and flexibility of DAM, 1D and 2D test cases are then considered. The results are discussed and compared with other available literature solutions.

2. The discrete ordinates equation

For an absorbing-emitting-scattering medium with a varying refractive index, the RTE in terms of the discrete coordinate s can be written as [25] :

$$n^2 \frac{d}{ds}\left[\frac{I(\Omega)}{n^2}\right] = -\beta I(\Omega) + n^2 \kappa I_b + \frac{\sigma_s}{4\pi}\int_{4\pi} I(\Omega')\Phi(\Omega',\Omega)d(\Omega') \tag{1}$$

where n is the refractive index; κ, σ_s and β are the coefficients of absorption, scattering and extinction, respectively; I_b is the blackbody intensity of the medium; and $\Phi(\Omega',\Omega)$ is the scattering phase function of the medium from direction Ω' to direction Ω. The single-scattering albedo is defined as $\omega = \sigma_s/\beta$.

The boundary conditions for diffuse walls can be written as

$$I(\Omega) = \varepsilon n^2 I_{bw} + \frac{1-\varepsilon}{\pi}\int_{\hat{n}\cdot\Omega'<0} I(\Omega')|\hat{n}\cdot\Omega'|d(\Omega'), \hat{n}\cdot\Omega > 0 \tag{2}$$

$$n^2 \frac{d}{ds}\left[\frac{I(\Omega)}{n^2}\right] = -\beta I(\Omega) + n^2 \kappa I_b + \frac{\sigma_s}{4\pi}\int_{4\pi} I(\Omega')\Phi(\Omega',\Omega)d(\Omega'), \hat{n}\cdot\Omega \le 0 \tag{3}$$

where ε is the wall emissivity, I_{bw} is the blackbody intensity of the wall, and $\hat{n}$ is the unit inward normal vector at the boundary location.

In a Cartesian coordinate system, the Eq. (1) can be written in its divergence form as described in [16] :

$$\begin{aligned}&\frac{dI(\Omega)}{ds} + \frac{1}{2n^2\sin\theta}\frac{\partial}{\partial\theta}\left\{I(\Omega)\left[\left(\xi\Omega - \vec{k}\right)\cdot\nabla n^2\right]\right\} + \frac{1}{2n^2\sin\theta}\frac{\partial}{\partial\varphi}\left\{I(\Omega)\left[\vec{s_1}\cdot\nabla n^2\right]\right\}\\ &= -\beta I(\Omega) + n^2\kappa I_b + \frac{\sigma_s}{4\pi}\int_{4\pi} I(\Omega')\Phi(\Omega',\Omega)d(\Omega')\end{aligned} \tag{4}$$

where

$$\Omega = \mu\cdot\vec{i} + \eta\cdot\vec{j} + \xi\cdot\vec{k} = \sin\theta\cos\varphi\cdot\vec{i} + \sin\theta\sin\varphi\cdot\vec{j} + \cos\theta\cdot\vec{k} \tag{5}$$

$$\vec{s_1} = -\sin\varphi\cdot\vec{i} + \cos\varphi\cdot\vec{j} \tag{6}$$

Using a piecewise constant angular (PCA) quadrature, the total solid angle is divided uniformly in N_θ polar and N_φ azimuthal directions. The discrete polar and azimuthal angles can be expressed as follows :

$$\theta^m = (m - 1/2)\Delta\theta,\ m = 1, 2, \ldots, N_\theta \tag{7}$$

$$\varphi^n = (n-1/2)\Delta\varphi,\ n=1, 2,\ldots, N_\varphi \tag{8}$$

where $\Delta\theta = \pi/N_\theta$, $\Delta\varphi = 2\pi/N_\varphi$.

For each discrete ordinate, the corresponding weight for θ^m and φ^n is

$$W_\theta^m = \int_{\theta^{m-1/2}}^{\theta^{m+1/2}} \sin\theta d\theta = \cos\theta^{m-1/2} - \cos\theta^{m+1/2} \tag{9}$$

$$W_\varphi^n = \int_{\varphi^{n-1/2}}^{\varphi^{n+1/2}} d\varphi = \varphi^{n+1/2} - \varphi^{n-1/2} \tag{10}$$

respectively, where $\theta^{m+1/2} = (\theta^m + \theta^{m+1})/2$, $\varphi^{n+1/2} = (\varphi^n + \varphi^{n+1})/2$.

Replacing the integral term by the discrete ordinates, yields to the discrete form [16] :

$$\frac{dI(\Omega^{m,n})}{ds} + \frac{1}{2n^2 \sin\theta}\frac{\partial}{\partial\theta}\left\{I(\Omega^{m,n})\left[\left(\xi\Omega^{m,n} - \vec{k}\right)\cdot\nabla n^2\right]\right\} + \frac{1}{2n^2\sin\theta}\frac{\partial}{\partial\varphi}\left\{I(\Omega^{m,n})\left[\vec{s}_1\cdot\nabla n^2\right]\right\}$$
$$= -\beta I(\Omega^{m,n}) + n^2\kappa I_b + \frac{\sigma_s}{4\pi}\sum_{m'=1}^{N_\theta}\sum_{n'=1}^{N_\varphi} I(\Omega^{m',n'})\Phi(\Omega^{m',n'},\Omega^{m,n})W_\theta^{m'}W_\varphi^{n'} \tag{11}$$

By discretizing the derivatives with respect to the polar θ and azimuthal φ angles, we obtain the discrete ordinate equation of the RTE in a 3D semitransparent refractive index media [16] :

$$\frac{dI(\Omega^{m,n})}{ds} + [\frac{1}{W_\theta^m}\max(\chi_\theta^{m+1/2,n},0) + \frac{1}{W_\theta^m}\max(-\chi_\theta^{m-1/2,n},0) + \frac{1}{W_\varphi^n}\max(\chi_\varphi^{m,n+1/2},0)$$
$$+\frac{1}{W_\varphi^n}\max(-\chi_\varphi^{m,n-1/2},0) + \beta]\cdot I(\Omega^{m,n}) = \frac{1}{W_\theta^m}\max(-\chi_\theta^{m+1/2,n},0)\cdot I(\Omega^{m+1,n})$$
$$+\frac{1}{W_\theta^m}\max(\chi_\theta^{m-1/2,n},0)\cdot I(\Omega^{m-1,n}) + \frac{1}{W_\varphi^n}\max(-\chi_\varphi^{m,n+1/2},0)\cdot I(\Omega^{m,n+1}) +$$
$$\frac{1}{W_\varphi^n}\max(\chi_\varphi^{m,n-1/2},0)\cdot I(\Omega^{m,n-1}) + n^2\kappa I_b + \frac{\sigma_s}{4\pi}\sum_{m'=1}^{N_\theta}\sum_{n'=1}^{N_\varphi} I(\Omega^{m',n'})\Phi(\Omega^{m',n'},\Omega^{m,n})W_\theta^{m'}W_\varphi^{n'} \tag{12}$$

where

$$\chi_\theta^{m+1/2,n} - \chi_\theta^{m-1/2,n} = \frac{W_\theta^m}{\sin\theta^m}\left[\cos 2\theta^m(\alpha'\cos\varphi^n + \beta'\sin\varphi^n) - \gamma'\sin 2\theta^m\right] \tag{13}$$

$$\chi_\theta^{1/2,n} = \chi_\theta^{N_\theta+1/2,n} = 0 \tag{14}$$

$$\chi_\varphi^{m,n+1/2} - \chi_\varphi^{m,n-1/2} = -\frac{W_\varphi^n}{\sin\theta^m}(\alpha'\cos\varphi^n + \beta'\sin\varphi^n) \tag{15}$$

$$\chi_\varphi^{m,1/2} = \chi_\varphi^{m,N_\varphi+1/2} = \frac{\beta'}{\sin\theta^m} \tag{16}$$

In the previous equations, α', β' and γ' correspond to the components of vector $(\alpha'\ \ \beta'\ \ \gamma') = \frac{1}{n}\overrightarrow{\mathrm{grad}}\, n$.

The boundary conditions in discrete form are

$$I(\Omega^{m,n}) = \varepsilon n^2 I_{bw} + \frac{1-\varepsilon}{\pi}\sum_{\hat{n}\cdot\Omega^{m',n'}<0} I(\Omega^{m',n'})\left|\hat{n}\cdot\Omega^{m',n'}\right| W_\theta^{m'}W_\varphi^{n'},\ \hat{n}\cdot\Omega^{m,n} > 0 \tag{17}$$

The incident radiation is given by

$$G = \int_{4\pi} I(\Omega) d\Omega = \sum_{m=1}^{N_\theta} \sum_{n=1}^{N_\varphi} I(\Omega^{m,n}) W_\theta^m W_\varphi^n \tag{18}$$

The radiative heat flux is

$$q_r = q_{rx}\vec{i} + q_{ry}\vec{j} + q_{rz}\vec{k} \tag{19}$$

where

$$q_{rx} = \int_{4\pi} \mu I(\Omega) d\Omega = \sum_{m=1}^{N_\theta} \sum_{n=1}^{N_\varphi} \mu^{m,n} I(\Omega^{m,n}) W_\theta^m W_\varphi^n \tag{20}$$

$$q_{ry} = \int_{4\pi} \eta I(\Omega) d\Omega = \sum_{m=1}^{N_\theta} \sum_{n=1}^{N_\varphi} \eta^{m,n} I(\Omega^{m,n}) W_\theta^m W_\varphi^n \tag{21}$$

$$q_{rz} = \int_{4\pi} \xi I(\Omega) d\Omega = \sum_{m=1}^{N_\theta} \sum_{n=1}^{N_\varphi} \xi^{m,n} I(\Omega^{m,n}) W_\theta^m W_\varphi^n \tag{22}$$

3. Energy equation

The steady state energy equation for coupled radiative and conductive heat transfer can be written as

$$k\Delta T - \nabla \cdot q_r = 0 \tag{23}$$

For a refractive index media, the divergence of the radiative flux is given by

$$\nabla \cdot q_r = \kappa\left(4n^2\sigma T^4 - G\right) \tag{24}$$

where G is the incident radiation. We define the conduction-radiation parameter $N = k\beta / 4\sigma T_{ref}^3$.

The conductive heat flux is

$$q_c = q_{cx}\vec{i} + q_{cy}\vec{j} + q_{cz}\vec{k} \tag{25}$$

where

$$q_{cx} = -k\frac{dT}{dx} \tag{26}$$

$$q_{cy} = -k\frac{dT}{dy} \tag{27}$$

$$q_{cz} = -k\frac{dT}{dz} \tag{28}$$

4. Diffuse approximation meshless method

Considering an unknown scalar u defined in a 2D field, at each nodal point X in the calculation domain, the second order Taylor expansion of u at the points X_i situated in the vicinity of X is estimated by [26] :

$$u_i^{estimated} = \langle P(X_i, X)\rangle \langle \alpha(X)\rangle^T \tag{29}$$

where $\langle P(X_i, X)\rangle$ is a vector of polynomial basis functions, defined as

$$\langle P(X_i, X)\rangle = \left\langle 1, (x_i - x), (y_i - y), (x_i - x)^2, (x_i - x)(y_i - y), (y_i - y)^2 \right\rangle \tag{30}$$

and $\langle \alpha(X)\rangle^T$ is a vector of coefficients, given by :

$$\langle \alpha_1, \alpha_2, \alpha_3, \alpha_4, \alpha_5, \alpha_6 \rangle^T = \langle u, \frac{\partial u}{\partial x}, \frac{\partial u}{\partial y}, \frac{1}{2!}\frac{\partial^2 u}{\partial x^2}, \frac{\partial^2 u}{\partial x \partial y}, \frac{1}{2!}\frac{\partial^2 u}{\partial y^2} \rangle^T \tag{31}$$

which are determined by minimizing the weighted discrete L_2 norm, defined as

$$I(\alpha) = \sum_{i=1}^{N} \omega_X(X_i, X)\left[u_i - \langle P(X_i, X)\rangle \langle \alpha(X)\rangle^T\right]^2 \tag{32}$$

in which $\omega_X(X_i, X)$ is the moving least-squares weight function. Its value decreases with increasing distance between nodal points of X_i and X, from unity at nodal point X to zero outside a given domain of influence.

The minimization of Eq. (32) gives:

$$A(X)\alpha(X) = B(X) \tag{33}$$

where

$$A(X) = \sum_{i=1}^{N} \omega_X(X_i, X) P(X_i, X) P^T(X_i, X) \tag{34}$$

$$B(X) = \sum_{i=1}^{N} \omega_X(X_i, X) P(X_i, X) u_i \tag{35}$$

By inverting Eq. (33), the components of $\langle \alpha(X) \rangle$ can be estimated by using the information of neighboring nodal points X_i :

$$\langle u, \frac{\partial u}{\partial x}, \frac{\partial u}{\partial y}, \frac{1}{2!}\frac{\partial^2 u}{\partial x^2}, \frac{\partial^2 u}{\partial x \partial y}, \frac{1}{2!}\frac{\partial^2 u}{\partial y^2} \rangle^{*T} = [A(X)]^{-1}\left\{\sum_{i=1}^{n(X)} \omega_X(X_i, X)\langle P(X_i, X)\rangle^T u_i\right\} \tag{36}$$

In this work, the Gaussian window:

$$\omega_X(X_i, X) = \exp\left[-3\ln 10\left(\frac{|X_i - X|}{\sigma_{\text{inf}}}\right)^2\right]$$
$$\omega_X(X_i, X) = 0 \quad \text{if } |X_i - X|^2 > \sigma_{\text{inf}}^2 \tag{37}$$

has been used in the calculations. $|X_i - X|$ defines the distance between nodal points of X_i and X. σ_{inf} is the distance of influence, which is used to select a certain number of neighboring nodal points $n(X)$ in order to preserve a local character of the approximation and also to ensure the matrix $A(X)$ is not singular. An initial value σ_{inf} is set first. Then, the distance $|X_i - X|$ is calculated. For 2D problems, if the number of nodal points that have a distance of $|X_i - X| < \sigma_{\text{inf}}$ is greater than 25 or smaller than 9, σ_{inf} is decreased or increased until $n(X)$ is just in the allowed range.

At each nodal point X in the computational spatial domain, the derivatives appearing in the equation (governing equations or boundary conditions) to be solved are replaced by their diffuse approximation in Eq. (36). As a result, an algebraic system is generated finally and can be solved by using the BICGSTAB iterative method.

5. Discretization schemes for governing equations and boundary conditions

In the following, the i th line of the inverse matrix $[A^M]^{-1}$ is defined as $\langle a_i \rangle$.

5.1. One dimensional problem

For 1D problem, the discrete ordinate equation of RTE (12) and boundary condition (3) (with $\hat{n}\cdot\Omega\leq 0$) can be written as :

$$\mu^m \frac{\partial I(\Omega^m)}{\partial x}+[\frac{1}{W_\theta^m}\max(\chi_\theta^{m+1/2},0)+\frac{1}{W_\theta^m}\max(-\chi_\theta^{m-1/2},0)+\beta]I(\Omega^m)$$
$$=\frac{1}{W_\theta^m}\max(-\chi_\theta^{m+1/2},0)I(\Omega^{m+1})+\frac{1}{W_\theta^m}\max(\chi_\theta^{m-1/2},0)I(\Omega^{m-1}) \quad (38)$$
$$+n^2\kappa I_b+\frac{\sigma_s}{2}\sum_{m'=1}^{N_\theta}I(\Omega^{m'})\Phi(\Omega^{m'},\Omega^m)W_\theta^{m'}$$

where $\mu^m=\cos\theta^m$ is the direction cosine related to the x-axis.

At the same time, Eq. (36) is transformed to :

$$\begin{bmatrix} u & \frac{\partial u}{\partial x} & \frac{1}{2!}\frac{\partial^2 u}{\partial x^2}\end{bmatrix}^{*T}=[A(X)]^{-1}\left\{\sum_{i=1}^{n(X)}\omega_X(X_i,X)\langle P(X_i,X)\rangle^T u_i\right\} \quad (39)$$

with

$$\langle P(X_i,X)\rangle=\langle 1,(x_i-x),(x_i-x)^2\rangle \quad (40)$$

By using the diffuse approximation in Eq. (39) at each nodal point X, the following algebraic systems for the radiative (RAD) and conductive (COND) equations are obtained respectively:

$$[RAD]\times[I]=[S_R] \quad (41)$$
$$[COND]\times[T]=[S_C] \quad (42)$$

where we have:

$$RAD(i,j)=\omega_X(X_j,X_i)\{\mu^m\langle a_2\rangle\}\langle P(X_j,X_i)\rangle^T+[\frac{1}{W_\theta^m}\max(\chi_\theta^{m+1/2},0)+\frac{1}{W_\theta^m}\max(-\chi_\theta^{m-1/2},0)$$
$$+\beta]\delta_{ij} \quad (43)$$

where δ_{ij} is Kroncker symbol.

$$COND(i,j)=\omega_X(X_j,X_i)\{2!k\langle a_3\rangle\}\langle P(X_j,X_i)\rangle^T \quad (44)$$

and

$$S_R(i)=\frac{1}{W_\theta^m}\max(-\chi_\theta^{m+1/2},0)I(\Omega^{m+1})+\frac{1}{W_\theta^m}\max(\chi_\theta^{m-1/2},0)I(\Omega^{m-1})$$
$$+n^2\kappa I_b+\frac{\sigma_s}{2}\sum_{m'=1}^{N_\theta}I(\Omega^{m'})\Phi(\Omega^{m'},\Omega^m)W_\theta^{m'} \quad (45)$$

$$S_C(i)=\kappa(4n^2\sigma T^4-G) \quad (46)$$

5.2. Two dimensional problem

In this case, the discrete ordinate equation of RTE (12) and boundary condition (3) (with $\hat{n}\cdot\Omega\leq 0$) can be written:

$$
\begin{aligned}
&\mu^{m,n}\frac{\partial I(\Omega^{m,n})}{\partial x}+\eta^{m,n}\frac{\partial I(\Omega^{m,n})}{\partial y}+[\frac{1}{W_\theta^m}\max(\chi_\theta^{m+1/2,n},0)+\frac{1}{W_\theta^m}\max(-\chi_\theta^{m-1/2,n},0)\\
&+\frac{1}{W_\varphi^n}\max(\chi_\varphi^{m,n+1/2},0)+\frac{1}{W_\varphi^n}\max(-\chi_\varphi^{m,n-1/2},0)+\beta]I(\Omega^{m,n})\\
&=\frac{1}{W_\theta^m}\max(-\chi_\theta^{m+1/2,n},0)I(\Omega^{m+1,n})+\frac{1}{W_\theta^m}\max(\chi_\theta^{m-1/2,n},0)I(\Omega^{m-1,n})\\
&+\frac{1}{W_\varphi^n}\max(-\chi_\varphi^{m,n+1/2},0)I(\Omega^{m,n+1})+\frac{1}{W_\varphi^n}\max(\chi_\varphi^{m,n-1/2},0)I(\Omega^{m,n-1})\\
&+n^2\kappa I_b+\frac{\sigma_s}{4\pi}\sum_{m'=1}^{N_\theta}\sum_{n'=1}^{N_\varphi}I(\Omega^{m',n'})\Phi(\Omega^{m',n'},\Omega^{m,n})W_\theta^{m'}W_\varphi^{n'}
\end{aligned}
\tag{47}
$$

where $\mu^{m,n}=\sin\theta^m\cos\varphi^n$ and $\eta^{m,n}=\sin\theta^m\sin\varphi^n$ are the two components of the direction vector $\Omega^{m,n}$.

The coefficients of the algebraic systems for RAD and COND equations are written as:

$$
\begin{aligned}
&RAD(i,j)=\omega_X(X_j,X_i)\{\mu^{m,n}\langle a_2\rangle+\eta^{m,n}\langle a_3\rangle\}\{P(X_j,X_i)\}^T+[\frac{1}{W_\theta^m}\max(\chi_\theta^{m+1/2,n},0)+\\
&\frac{1}{W_\theta^m}\max(-\chi_\theta^{m-1/2,n},0)+\frac{1}{W_\varphi^n}\max(\chi_\varphi^{m,n+1/2},0)+\frac{1}{W_\varphi^n}\max(-\chi_\varphi^{m,n-1/2},0)+\beta]\delta_{ij}
\end{aligned}
\tag{48}
$$

$$
COND(i,j)=\omega_X(X_j,X_i)\cdot[2!k(\langle a_4\rangle+2!\langle a_6\rangle)]\cdot\langle p_j\rangle^T \tag{49}
$$

and

$$
\begin{aligned}
S_R(i)&=\frac{1}{W_\theta^m}\max(-\chi_\theta^{m+1/2,n},0)I(\Omega^{m+1,n})+\frac{1}{W_\theta^m}\max(\chi_\theta^{m-1/2,n},0)I(\Omega^{m-1,n})\\
&+\frac{1}{W_\varphi^n}\max(-\chi_\varphi^{m,n+1/2},0)I(\Omega^{m,n+1})+\frac{1}{W_\varphi^n}\max(\chi_\varphi^{m,n-1/2},0)I(\Omega^{m,n-1})\\
&+n^2\kappa I_b+\frac{\sigma_s}{4\pi}\sum_{m'=1}^{N_\theta}\sum_{n'=1}^{N_\varphi}I(\Omega^{m',n'})\Phi(\Omega^{m',n'},\Omega^{m,n})W_\theta^{m'}W_\varphi^{n'}
\end{aligned}
\tag{50}
$$

$$
S_C(i)=\kappa(4n^2\sigma T^4-G) \tag{51}
$$

5.3. Boundary conditions

For nodal points lying on the boundary with $\hat{n}\cdot\Omega^{m,n}>0$, the following coefficients for Eq. (17) in 1D and 2D are obtained respectively:

$$
RAD(i,j)=\delta_{ij} \tag{52}
$$

$$
S_R(i)=\varepsilon n^2 I_{bw}+2(1-\varepsilon)\sum_{\hat{n}\cdot\Omega^{m'}<0}I(\Omega^{m'})|\hat{n}\cdot\Omega^{m'}|W_\theta^{m'}\ ,1\text{D} \tag{53}
$$

$$
S_R(i)=\varepsilon n^2 I_{bw}+\frac{1-\varepsilon}{\pi}\sum_{\hat{n}\cdot\Omega^{m',n'}<0}I(\Omega^{m',n'})|\hat{n}\cdot\Omega^{m',n'}|W_\theta^{m'}W_\varphi^{n'}\ ,2\text{D} \tag{54}
$$

The Dirichlet boundary conditions for energy equation are introduced in the same way as Eq. (17).

5.4. Iterative process

The following iterative process is chosen to solve the non linear coupled systems.

1. A finite number of nodal points in the domain and on the boundaries are chosen first. Then, the derivatives of all the points are replaced by their diffuse approximation.
2. Initial temperature field is set.
3. The discrete ordinate equation of RTE is solved.
4. Then the energy equation is solved.
5. Terminate the iteration process if the convergence criterion is satisfied. Otherwise, go back to step 3.

The convergence criterion used for the numerical calculation is defined as $\left|\frac{T_{new}-T_{old}}{T_{new}}\right|_{\max} \le 10^{-5}$.

6. Results

To facilitate comparison of the results obtained with DAM and those obtained in the literature, the relative error is defined as :

$$E_r = \frac{\sqrt{\sum\left(T^{num}-T^{ref}\right)^2}}{\sqrt{\sum\left(T^{ref}\right)^2}} \times 100\% \tag{55}$$

6.1. Plane parallel slab

This case considers an absorbing and emitting medium contained in the plane parallel slab with a thickness L=0.01 m. The temperatures of left and right walls are T_0=1000 K and T_L=1500 K, respectively.

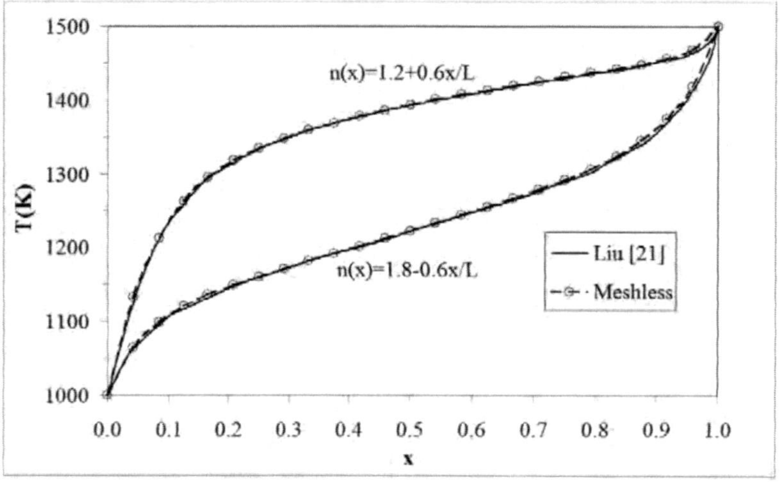

Figure 1. Temperature distributions across the slab for different linear refractive index, $\varepsilon_0=\varepsilon_L=1$

The temperature distributions across the slab are calculated by using a discretization of 21 nodal points and N_θ=40 quadrature. The coefficient of absorption is κ=100.0 m^{-1} (optical thickness τ=1) and the thermal conductivity is k=0.1 W m^{-1} K^{-1}. Firstly we consider the case of black surfaces ($\varepsilon_0=\varepsilon_L$=1) with different linear refractive index, namely, $n(x)$=1.2+0.6x/L and $n(x)$=1.8-0.6x/L. The influence of refractive index $n(x)$ on the temperature distribution is shown in Figure 1. It can be seen that when the gradient of refractive index is positive, corresponds to $n(x)$=1.2+0.6x/L, the temperature is globally higher. Further, the influence of emissivity on the temperature distribution is presented in Figure 2 for a sinusoidal refractive index $n(x)$=1.8-0.6$\sin(\pi x/L)$.When the emissivities of the two walls are not equal, the distributions of temperature are greatly changed comparing with the situation of two black walls. The maximum relative error for the temperature distribution is 0.5721%.

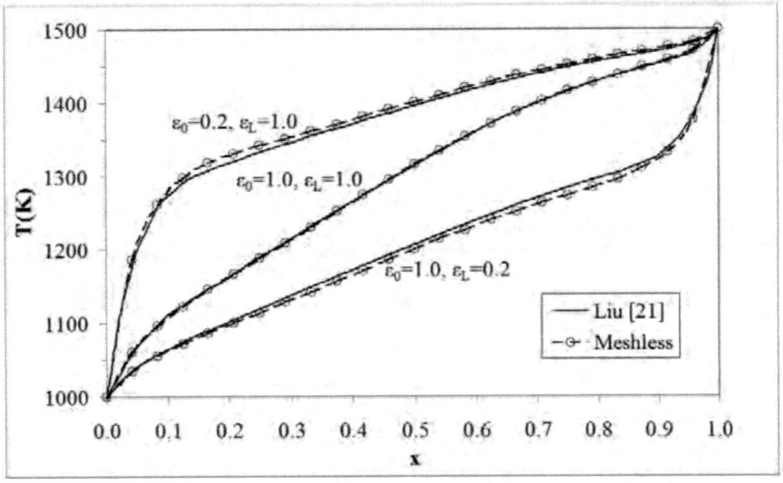

Figure 2. Temperature distributions across the slab for different emissivity, $n(x)$=1.8-0.6 $\sin(\pi x/L)$

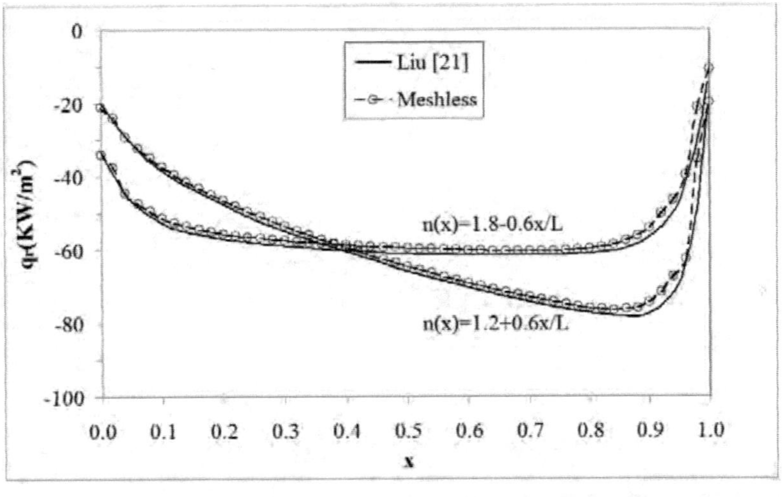

Figure 3. Radiative heat flux distributions across the slab for different linear refractive index, ε_0=1.0, ε_L=0.2

The radiative heat flux distributions across the slab are calculated for a discretization of 51 nodal points and N_θ=80 quadrature. The coefficient of absorption is κ=1000.0 m^{-1} (optical thickness τ=10) and the thermal conductivity is k=1.0 W m^{-1} K^{-1}. Firstly we consider linear refractive index with the same expression $n(x)$=1.2+0.6 x/L and $n(x)$=1.8-0.6 x/L as before but with different emissivities ε_0=1.0 and ε_L=0.2 on the two walls. The influence of refractive index $n(x)$ on the temperature distribution is shown in Figure 3. Near the left wall with ε_0=1.0, once again, the radiative heat flux is much higher for refractive index whose gradient is positive; but the phenomenon is on the contrary near the right wall. Next, we consider the case of black surfaces with different sinusoidal refractive index, namely, $n(x)$=1.8-0.6 $\sin(\pi x/L)$ and $n(x)$=1.2+0.6 $\sin(\pi x/L)$. The radiative heat flux is shown in Figure 4 and presents a kind of particular symmetry in function of refractive index. The maximum relative error for the radiative heat flux is 2.1973%.

All the results have a good agreement with those by Liu [21] who employed MLPG approach.

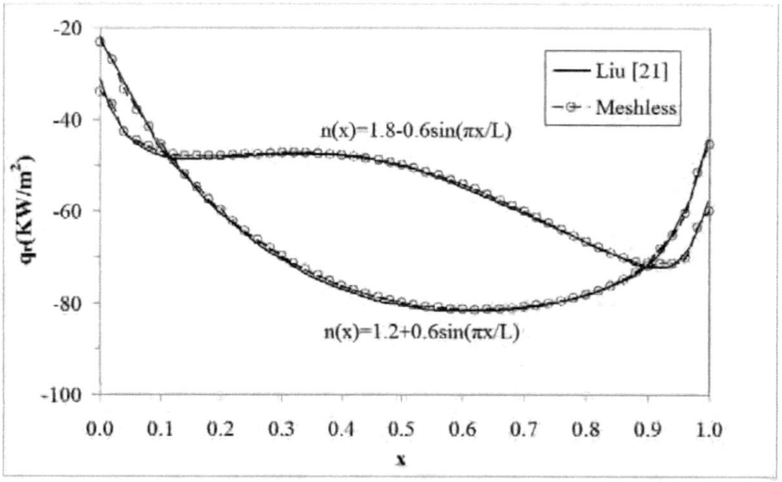

Figure 4. Radiative heat flux distributions across the slab for different sinusoidal refractive index, $\varepsilon_0 = \varepsilon_L = 1$

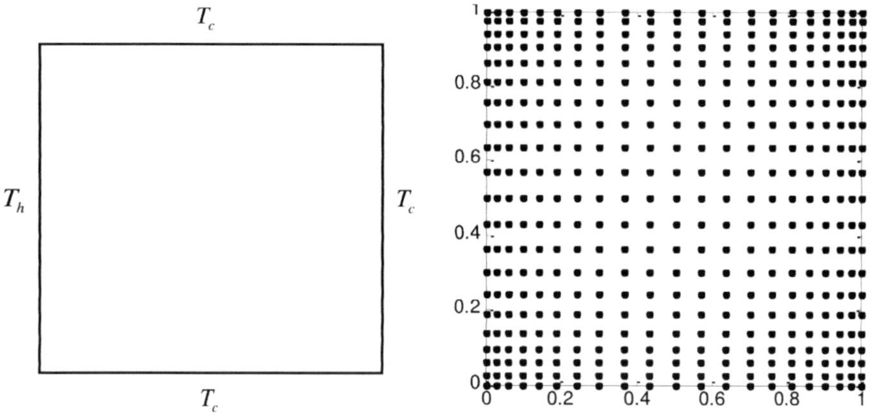

Figure 5. Square enclosure and nodal points

6.2. Square

We consider an absorbing and emitting medium contained in a square as shown in Figure 5. All the walls are black. The left one is hot at temperature T_h and the other three walls are cold at temperature T_c, such that T_h/T_c =0.5. The absorption coefficient and conduction-radiation parameter are set to κ =1.0 m^{-1} and N =0.01 respectively. We use a discretization of 21×21 nodal points and $N_\theta \times N_\varphi$=20×40 quadrature.

The temperature distributions in refractive index media with linear variation along the two directions x and y are shown in Figure 6 for a gradient value of 0, 1, and 2, respectively. We can find that the refractive index $n(x,y)$=1 conducts the maximum temperature distribution and the temperature decreases as the gradient value of refractive index increases. We can see the results obtained by DAM agree well with those in reference [18] by a finite element method with a maximum error of 0.8524%.

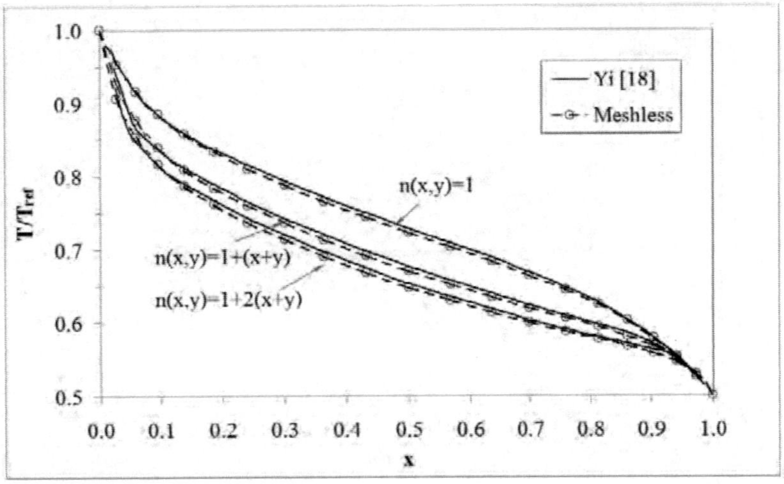

Figure 6. Distribution of the temperature along centerline y =0.5 for different linear refractive index (κ =1.0 m^{-1}, ε =1, N =0.01)

6.3. Semicircular enclosure

A semicircular enclosure with inner circle is shown in Figure 7 and all the walls are black. The same spatial discretization of 1144 nodal points is used for the spatial discretizaiton. For the angular discretization, different from former cases in which S_N method is used, here we use a PCA (piecewise constant angular quadrature) method and a number of discrete directions $M_\theta \times M_\varphi = 20 \times 40$ is used. The medium is absorbing, emitting and scattering. The temperature of the outside semicircle and inner circle are T_s =1000K and T_i =500K, respectively. We use the mixed boundary condition on the bottom as [27]:

$$-k\frac{\partial T}{\partial \hat{n}_w} + \sum_{m=1}^{N_\theta}\sum_{n=1}^{N_\varphi} I\left(\Omega^{m,n}\right)\left(\Omega^{m,n}\cdot\hat{n}_w\right)W_\theta^m W_\varphi^n = h\left(T - T_f\right) + \varepsilon\sigma\left(T^4 - T_\infty^4\right) \tag{56}$$

where h =20 W m^2 k^{-1}, $T_f = T_\infty$ =300 K.

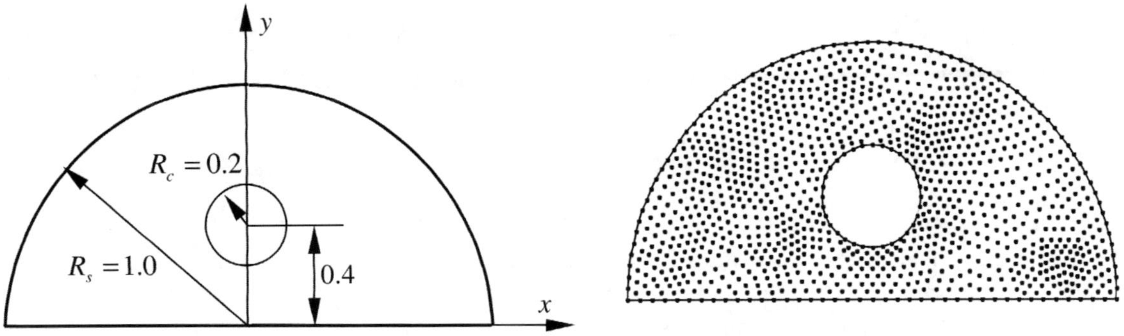

Figure 7. Semicircular enclosure with inner circle and the distribution of nodal points

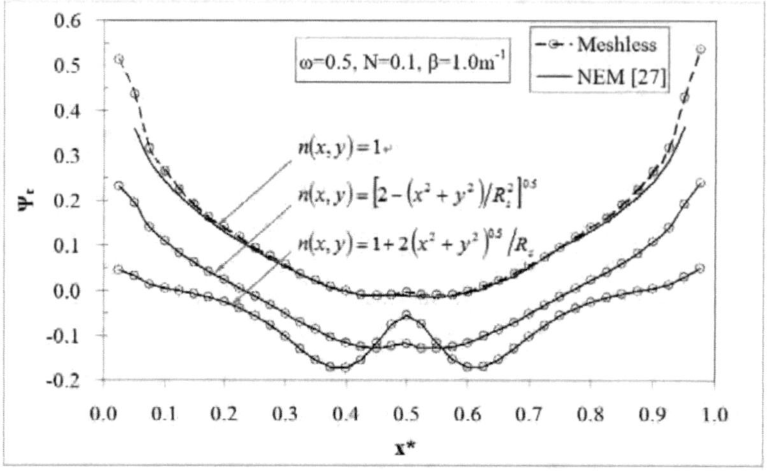

Figure 8. Effects of various refractive index on the dimensionless conductive heat flux Ψ_c along the bottom wall

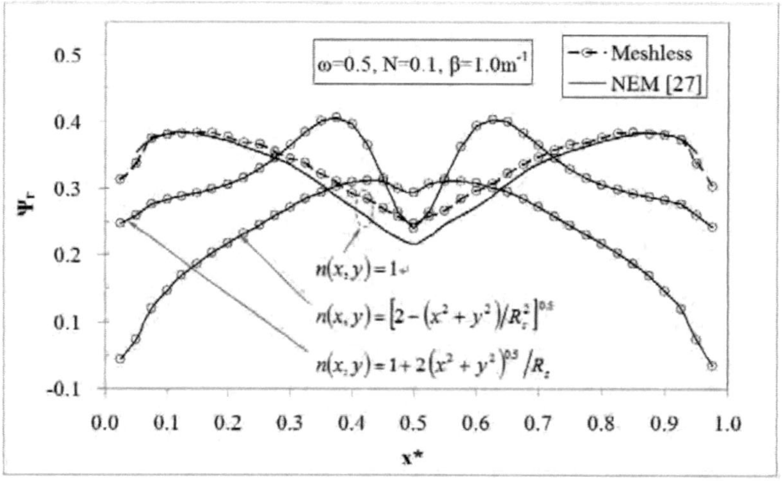

Figure 9. Effects of various refractive index on the dimensionless radiative heat flux Ψ_r along the bottom wall

For β=1.0 m^{-1}, N=0.1 and ω=0.5, the dimensionless radiative heat flux $\Psi_r = q_r/\sigma T_s^4$ and conductive heat flux $\Psi_c = q_c/\sigma T_s^4$ for different kinds of refractive index $n(x,y)$ =1, $\left[2-\left(x^2+y^2\right)/R_s^2\right]^{0.5}$ and $1+2\left(x^2+y^2\right)^{0.5}/R_s$ are shown in Figure 8 and Figure 9. For $n(x)$=1, our results are compared with those reported in reference [27,28]. Although the relative error E_r is significant and equals to 5.7698% and 11.5% for radiative and conductive heat flux respectively, the trend is similar. Besides, there are not available results in other references for the problem. For all kinds of refractive index, the dimensionless conductive heat flux Ψ_c distributions present a symmetrical characteristic, which reach the peak in the center of the bottom wall but the trough around the center. The trend is on the contrary for dimensionless radiative heat flux Ψ_r. The peak and the trough become obvious for refractive index $1+2\left(x^2+y^2\right)^{0.5}/R_s$, which increases with increasing distance from the center of outer circle. In Table 1, the CPU time necessary for convergence is also given for various refractive index. All the results have been obtained on a personal computer (3.1 GHz Processor, 8 GB RAM, without parallelization).

Table 1. Time necessary for different refractive index

$n(x,y)$	time
1	57 mins and 43 secs
$1+2(x^2+y^2)^{0.5}/R_s$	16 mins and 58 secs
$[2-(x^2+y^2)/R_s^2]^{0.5}$	29 mins and 58 secs

7. Conclusion

A diffuse approximation meshless method (DAM) based on the DOM of RTE is employed for solving radiative-conductive heat transfer in refractive index media contained in the 1D and 2D geometries. The results are compared with other benchmark results, and show that DAM has a good accuracy in solving radiative-conductive heat transfer in absorbing, emitting and scattering media with variation of refractive index in multidimensional geometries. The influence of the refractive index are investigated on the temperature and heat flux distributions. The future work will consider to extend the case to complex 3D geometries and also to improve the calculation efficiency.

8. References

[1] Boutina L and Bessaïh R 2011 *Appl. Therm. Eng.* **31** 2052-62
[2] Wang F Q, Shuai Y, Tan H P and Yu C L 2013 *Int. J. Heat Mass Tran.* **62** 247–54
[3] Sakami M, Charette A and Le Dez V 1996 *J. Quant. Spectrosc. & Radia. Tran.* **56** 517–33
[4] Ruan L M, Xie M, Qi H, An W and Tan H P 2006 *J. Quant. Spectrosc. & Radia. Tran.* **102** 190–202
[5] Talukdar P, Issendorff F V, Trimis D and Simonson C J 2008 *Heat Mass Tran.* **44** 695–704
[6] Ben Abdallah P and Le Dez V 2000 *J. Quant. Spectrosc. & Radia. Tran.* **67** 125–37
[7] Liu L H and Tan H P 2004 *J. Quant. Spectrosc. & Radia. Tran.* **83** 333–44
[8] Liu L H 2006 J. Quant. Spectrosc. & Radia. Tran. 102 293–303
[9] Xia X L, Huang Y, Tan H P and Zhang X B 2002 *Int. J. Heat Mass Tran.* **45** 2673–88
[10] Yi H L, Tan H P, Luo J F and Dong S K 2005 *J. Quant. Spectrosc. & Radia. Tran.* **96** 363–81
[11] Yi H L, Tan H P, Zhang H C and Luo J F 2006 *Numer. Heat Tran.* A **49** 607–34
[12] Yi H L, Xie M and Tan H P 2008 *Int. J. Heat Mass Tran.* **51** 5918–30
[13] Mishra S C, Krishna N A, Gupta N and Chaitanya G R 2008 *Int. J. Heat Mass Tran.* **51** 83–90
[14] Zhu K Y, Huang Y and Wang J 2013 Int. J. Heat Mass Tran. 62 200-04
[15] Lemonnier D and Le Dez V 2002 *J. Quant. Spectrosc. & Radia. Tran.* **73** 195–204
[16] Liu L H 2006 *J. Thermophys. Heat Tran.* **20** 59–66
[17] Liu L H, Zhang L and Tan H P 2006 *J. Quant. Spectrosc. & Radia. Tran.* **97** 446–56
[18] Yi H L, Zhen B, He K, Tan H P and Zhou Y 2010 *Kung Cheng Je Wu Li Hsueh Pao/J. Eng. Thermophys.* **31** 1941–44
[19] Sun Y S and Li B W 2010 *Int. J. Heat Mass Tran.* **53** 1491–97
[20] Zhang Y, Yi H L and Tan H P 2013 *Int. J. Heat Mass Tran.* **66** 349–54
[21] Liu L H, Tan J Y and Li B X 2006 *J. Quant. Spectrosc. & Radia. Trans.* **101** 237–48
[22] Nayroles B, Touzot G and Villon P 1991 *C. R. Acad. Sci. Paris* **313** 293–96
[23] Wang C A, Sadat H and Le Dez V 2012 *Appl. Math. Model.* **36** 5309−19
[24] Sadat H, Wang C A and Le Dez V 2012 *Appl.Math. Comput.* **218** 10211–25

[25] Preisendorfer R W 1965 *International Series of Monographs on pure and applied Mathematics* Pergamon Press

[26] Prax C, Sadat H and Dabboura E 2007 *Appl. Math. Comput.* **186** 1040–53

[27] Zhang Y, Yi H L and Tan H P 2013 *J. Comput. Phys.* **241** 18–34

[28] Wang C A, Sadat H and Tan J Y 2014 *Numer. Heat Tran.* B **65** 518–36

Acknowledgments

This work was supported by natural scientific research innovation foundation in Harbin Institute of Technology (HIT.NSRIF.2011110, HIT.NSRIF.2014132) scientific research foundation of Harbin Institute of Technology at Weihai (HIT(WH)201302), and Program for New Century Excellent Talents in University (NCET-12-0152).